T0139151

A K Peters

Editorial, Sales, and Customer Service Office
A K Peters, Ltd.
63 South Avenue
Natick, MA 01760

Library of Congress Cataloging-in-Publication Data

Schraft, R. D. (Rolf-Dieter), 1942-
[Serviceroboter. eng]
Service robots / Rolf Dieter Schraft, Gernot Schmierer.
p. cm.
Translation of Serviceroboter
ISBN 1-56881-109-8 (alk. paper)
1. Robotics. 2. Robots, Industrial. I. Schmierer, Gernot, 1969- II. Title.

TJ211 .S42 2000
629.8'92--dc21 00-024963

Authors' address: Professor Dr.-Ing. Dr. h. c. mult. Rolf D. Schraft
Dipl.-Ing. Gernot R. Schmierer
Fraunhofer-Institut für Produktionstechnik und Automatisierung
Nobelstr. 12
D-70569 Stuttgart (Germany)

Design by reform design Stuttgart
www.reform-design.de
zentrale@reform-design.de

Printed in Germany
03 02 01 00 10 9 8 7 6 5 4 3 2 1

Rolf Dieter Schraft

Gernot Schmierer

Service Robots

Products

Scenarios

Visions

Industrial Robots are Gaining Popularity

According to the last count of industrial robots by the United Nations Economic Commission for Europe, one million industrial robots will soon be operating in automated production throughout the world. Among the various industries now using industrial robots, the automobile industry is in the lead, where most of the industrial robots carry out welding tasks such as point welding.

It goes without saying that in industrial manufacturing, automated installations and systems are used to perform heavy, dirty, and repetitive tasks. In a few years' time, it is possible that robotic systems will also become a common sight in the service sector. Automating specialists in this field have focused on cleaning applications, in an effort to save humans from the unpleasant and sometimes dangerous task of industrial cleaning.

Technical advances in the fields of sensor, control, and drive technology enabled intelligent robotic systems to be implemented in areas other than industrial production.

The disproportionate growth in the service sector, in comparison with other branches, has made it necessary to consider optimization and improvements in work methods. Already, nearly all areas of the expanding service industry are introducing modern information and communication technology to carry out tasks economically and in a customer-friendly manner.

Further economic considerations and efforts for humane working conditions in the service sector increasingly point toward having partially or fully automated mechanized work sequences in the future.

In the mid- and long-term, leading robotics experts forecast that the number of service robots will overtake the number of industrial robots. Development is accelerating in this area. Innovations and, in particular, creativity, are the driving forces that give this process the required impulsion.

Together with Fraunhofer IPA, both of the editors of this book have taken on the important task of clarifying current developments. In their work, they have aimed to do this not only for the German research and development market, but in particular, to portray the present situation worldwide. This is an ongoing task. I wish them the strength and the stamina required to continue their efforts.

Prof. Dr.-Ing. Dr. h.c. mult. Hans-Jürgen Warnecke,
President of the Fraunhofer-Society

Innovation Incentive to the Service Sector

Since the beginning of the 1990s, the globalization process in Germany has accelerated. One reason for this is the addition of new market providers from the American and Asian regions, as well as the former East Block countries. The development has been markedly influenced by the information and communication technologies available today, which have made the previous problem of distance almost negligible.

The resurgence of competitiveness and thus a lasting revival of the German economy can, in view of this background, only be attained if we meet this global challenge with a powerful offer at suitable prices.

Various companies throughout the world have already recognized the march of progress and are in the process of adjusting their structures for the new conditions; by using innovative products and methods, they are creating room to move in their widened field of competition again.

With a view to the increasing importance of outsourcing, especially in service, the lasting strength of one's own competitive position will depend greatly on not only placing innovative impetus to the industrial sector, but particularly on steering towards the tertiary area. In this way, and with a good value-to-money ratio on the offer, the goal must be to increase the utilization of existing services and to open up new fields of operation for individuals in services.

Service robots are particularly suited to this requirement, and the authors of this book have shown this effectively in many examples. In this context, modern technology will open up promising development perspectives and, to some extent, even create completely new fields of application for the "service society" of tomorrow.

Here, not only existing companies but also new ones and spin-offs will have an excellent chance to participate in this process if they have innovative concepts. New companies have several advantages. As well as possessing creativity and the ability to implement changes, new companies also are free from the bureaucratic structure of large companies, and therefore are able to be flexible and quick-moving in order to keep up with the market. In our experience, the most promising companies are those that put their concepts into action using strategic alliances, and optimize their company performance in this respect with professional coaching. On the basis of such partnerships, (e.g., with "venture capital" societies or established companies) the IKB Deutsche Industriebank AG supports innovative projects with equity and borrowed capital, and therefore contributes toward anchoring new technologies in the German industrial and service sectors to bring about an urgently necessary increase in productivity.

Georg-Jesko v. Puttkamer,
Member of the Board of Directors,
IKB Deutsche Industriebank AG

The Evolution from the Industrial Robot to the Personal Robot

The robot made its final breakthrough into industrial manufacturing in the 1980s. This trend for industrial robots is increasing in the 1990s. Statisticians forecast that by the turn of the millennium, over 800.000 industrial robots will be installed worldwide.

Despite this, renowned robot manufacturers are already changing direction: Due to the foreseeable saturation of the market, they are investing in the service sector as a future field of application with enormous growth potential. Innovative developments in the area of sensor, control, and drive technologies are opening up a whole variety of new implementation possibilities for robotic systems outside of the field of industrial production. In the evolution from industrial to personal robots, these so-called service robots represent the halfway point. They are mobile, manipulate things, interact with people, and carry out tasks independently to relieve the strain on humans. Their partially or fully automated tasks do not serve the industrial manufacture of goods, but instead, perform tasks for people and equipment: They perform services.

The disproportionate growth in the service sector, in comparison with the primary and secondary sectors, is a widely accepted fact. Growth increases in the area of service are seen as marks of progress; the spectrum and availability of services is an important measurement of quality-of-life.

The use of service robots in the future provides the opportunity to perform tasks in the service sector

- economically,
- dependably,
- flexibly and individually,
- at a qualitatively high standard, and
- in an ecologically and socially acceptable manner.

In numerous service sectors, efforts are being taken either to develop service robots or to implement them on a greater scale. In contrast with the field of industrial robots, the majority of service robots will be adjusted individually according to their type, environment, and task sequence, even if important subsystems such as mobile platforms, drive and control techniques, sensors, and operating systems are constructed in a modular form and can be applied to other operational uses.

This book does not claim to be comprehensive, but rather tries to give a representative overall view of the developments in the field of service robots worldwide. In many countries at the moment, various research and development programs are running on a broad scope-of-field. Among them are large research projects financed by the state, as well as discoveries made by enthusiastic engineers; they all see a chance to find a place for their product in the growing commercial market.

We have not only set a precedent in the area of visual and graphical representation with this book, we also present a direction for the future in the field of marketing. Thanks to Mr. Hubert Grosser, personal assistant to the President of the Fraunhofer Society, we have found a new partner in the IKB Deutsche Industriebank AG, which has already been involved with innovative strategies and visions in the past.

We would also like to thank the employees of the Fraunhofer IPA, who have played a decisive role in this book with their technical contributions and critical discussions. In particular, we would like to mention Mrs. Andrea Hiller and also Mr. Baum, Mr. Cottone, Mr. Dahlkemper, Mr. Erhardt, Mr. Gehringer, Mr. Hägele, Mr. Hornemann, Mr. Meißner, Mr. Müller, Mr. Ritter, Mr. Rust and Mr. Andreas Wolf, who, in addition, encouraged us to use a new graphical layout.

We would like to thank Mrs. Melissa Delein for her remarkable commitment to data research. The technical journalistic support given by Mr. Roland Dreyer helped us in our work enormously. We would also like to thank Mrs. Helen Schliesser for translating the German edition of this book with a dedication to perfection. Mr. Kellner, Mr. Kotulla and Mr. Kull designed the graphical layout with considerable dedication and thought to giving a suitable form to the innovative contents of this book.

Last but not least we would like to thank Alice and Klaus Peters, Kathryn Maier, and the production team at AK Peters.

Rolf Dieter Schraft, Gernot Schmierer
Stuttgart, Germany, December 1999

WAIT
open>

Contents

Service Robots Are Key Products Showing the Way for the Future

Service robots refuel vehicles, renovate nuclear power stations, care for the elderly, keep watch over museums, explore Mars, and clean airplanes. So what is a service robot?

To date, various institutions and organizations are trying to find a tangible definition. The International Federation of Robotics (IFR) developed a suggestion in 1997:

"A service robot is a robot that operates partially or fully autonomously to perform services useful to the well-being ... of humans and equipment. They are mobile or manipulative or a combination of both."

This statement is based on the definition suggested by the Fraunhofer Institute for Manufacturing Engineering and Automation (IPA) in 1994:

"A service robot is a freely programmable mobile device carrying out services either partially or fully automatically. Services are tasks that do not serve the direct industrial manufacture of goods, but the performing of services for humans and equipment."

Certainly neither definition is all-inclusive. One thing is clear, though: The areas of application of service robots are not easily defined as were those of the industrial robot.

Robot Manufactures are Looking for New Markets

Industrial robots were first introduced into factories in the 1960s, and the number of installed units grew to about 860,000 by 1996. However, because many industrial robots of the first generation have since been scrapped, the United Nations European Commission for the Economy (UN/ECE) and the IFR assume that in 1998 approximately 720,000 industrial robots were performing their duties worldwide; that is six percent more than in 1996.

With over 400,000 systems, more than half of these robots have been installed in Japan. Germany, with about 60,000 industrial robots, lies in third place, well behind the USA. The study carried out by the UN/ECE and the IFR predicts that up to 800,000 industrial robots will be in use worldwide by the year 2000.

Expressed in percentages, the highest increase will take place in Asian countries (not including Japan), according to this estimate. The German market is relatively stable at the moment. However, renowned robot manufacturers are already looking for new markets that will guarantee a high turnover in the future. Their attention is turning more and more towards the service sector, which will have enormous growth potential in the future.

Service Robots Have Growth Potential

A service robot in every household? A service robot as a part of the basket of available commodities in 2015? Service robots as mass products of the twenty-first century? These questions concern engineers throughout the world who have devoted their work to this fascinating field.

The examples listed at the beginning of this chapter show that it is almost impossible to give global potential estimates for the use of service robots. Due to the comparatively high diversity and the small number of service robots installed to date, statistics, such as they exist in the area of industrial robots, have only been recorded occasionally.

If one sees the service robot as a halfway step in the evolution of industrial robots towards personal robots, then this product indeed has all the prerequisites for becoming a key product in the near future.

Just like the locomotive, electrical engineering, the automobile, plastics, and, finally, television and mobile phones, the service robot could establish itself in the mainstream of information and communication technologies as a mass product by the beginning of the twenty-first century.

A Book for Practical Use

This book will accompany the inventor or developer of a service robot from the idea to the point of creation. By presenting exemplary solutions that have already been realized as well as prototypes from applied research, it may also inspire others to design innovative service robots. At the end, exemplary visions will also be put forward; they may perhaps soon mature into research projects.

From Idea to Product

The Fusion of Software and Mechanics

Service robots are constructed by a synthesis of mechanics, control techniques, and software. Their development is a complex and sensitive process. Here, we show the systematic sequence of phases that comprise the steps from the idea to the product.

For this, we have used methods that have already been successfully implemented and refined in the development of service robots. The interweaving of technical and economical aspects is crucial in the planning phase. Today, equal amounts of technical competence and economical thinking are expected of a design engineer.

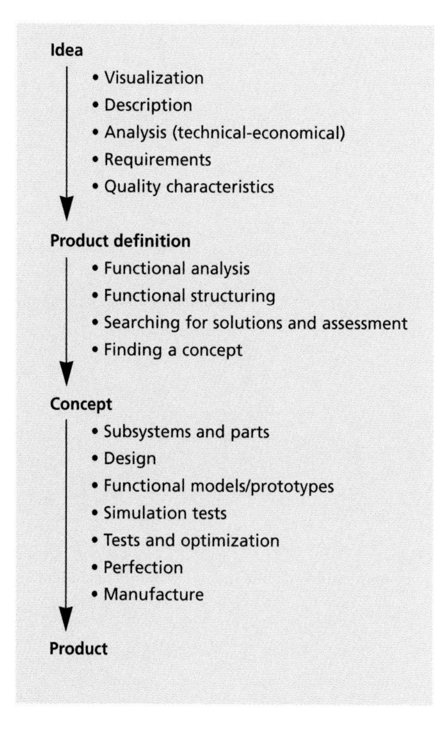

Idea
• Visualization
• Description
• Analysis (technical-economical)
• Requirements
• Quality characteristics
Product definition
• Functional analysis
• Functional structuring
• Searching for solutions and assessment
• Finding a concept
Concept
• Subsystems and parts
• Design
• Functional models/prototypes
• Simulation tests
• Tests and optimization
• Perfection
• Manufacture
Product

Economical and Technical Assessment Go Hand in Hand

The path follows the steps of the idea, to the definition of the product, the concept, and finally to product design. Each objective stage is reached primarily by taking the same elementary steps:

- Obtain information
- Search for a solution
- Analyze
- Assess
- Select

With this systematic sequence, the design engineer always has a clear basis for his decision whether to continue or to abort the project.

I.) From the Idea to the Definition of the Product

Development begins with the idea of a service robot, which is related either to the function or to the structure of the device. The robot's main tasks, the environment where it will be used, and the system use profile must be determined in order to define the idea.

The first step in defining the idea includes identifying the main function of the service robot (e.g., "cleaning floors"). The environment in which it will be implemented is then defined. This includes the physical place where it will be used, and the people who will provide, operate, and use the service robot. These three groups of people may or may not overlap.

providers
Providers are people or companies who represent the potential buyers of the service robot and who offer its services to a third party.

operators
Operators are responsible for handling and controlling the system. They also define the qualification requirements that the system has to meet in order to be functional.

users
Users utilize the services offered by the robot.

The System Use Profile Assesses Advantages

The system use profile enables a person to rate and assess the advantages of using a service robot. The profile is divided into financial, qualitative, socio-ethical, and image-related aspects. The profile permits a comparison to be made with similar existing services.

Should implementing a service robot be cost-effective, an estimated cost comparison on the basis of market potential and performance time is recommended.

Visualization Helps to Communicate

If an idea is to become reality, one needs to be able to vividly imagine and visualize it in all its detail. In addition to still diagrams and sketches, computer-based animation is the best way to visually represent an idea. Expenditure should be kept as low as possible at this largely uncertain stage of development. Very realistic animations can be produced with various software tools that are inexpensive and easy to use.

With visualization, ideas can be communicated much better and much more easily. The visualization conveys how a particular function should be carried out with which basic type of system. It also can give the design engineer a feel for the critical subsystems and functions.

Technical-Economical Analysis and Assessment

The technical analysis determines the principal technical feasibility. Based on a description of the idea, the critical subsystems and functions are identified, and their technical aspects and cost projections are researched.

For each critical subsystem, a realistic solution must be found from the technical and financial points of view. This is an essential prerequisite for development to continue. Development and manufacturing costs for the service systems then need to be estimated.

In the economical analysis, statistical data is used to determine the highest possible market potential for robot's services. The number of service systems required to perform the services defined in the market potential is then estimated. Next, the market potential for the service systems is divided up into target markets and target customers. By taking into consideration such factors as the market penetration goal (dependent on piece prices), the market potential can thus be assessed.

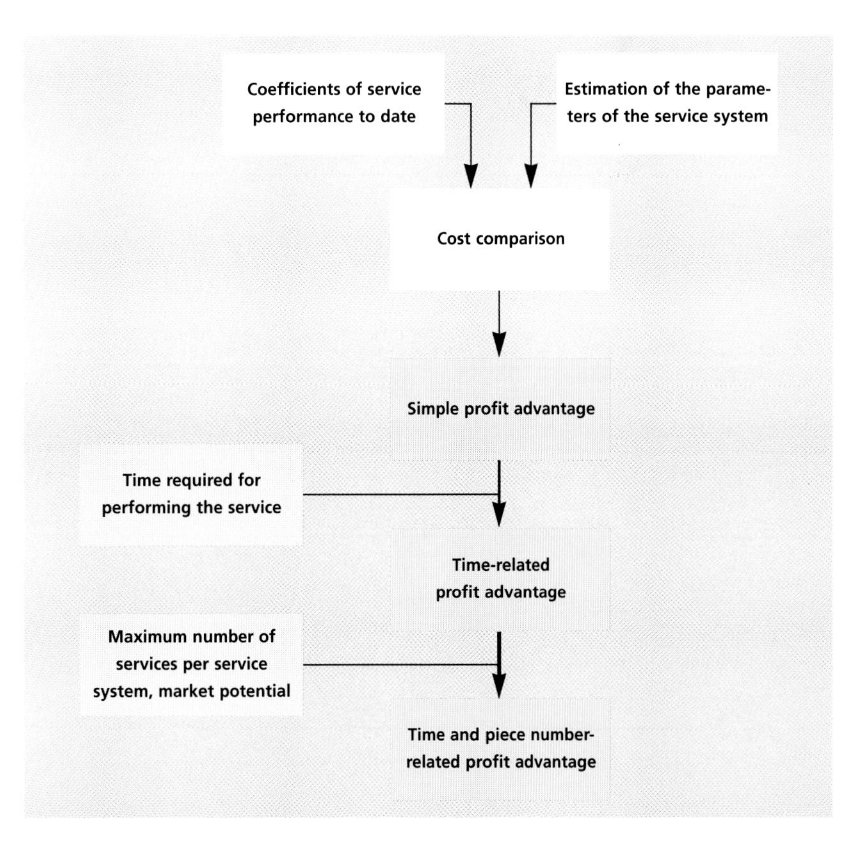

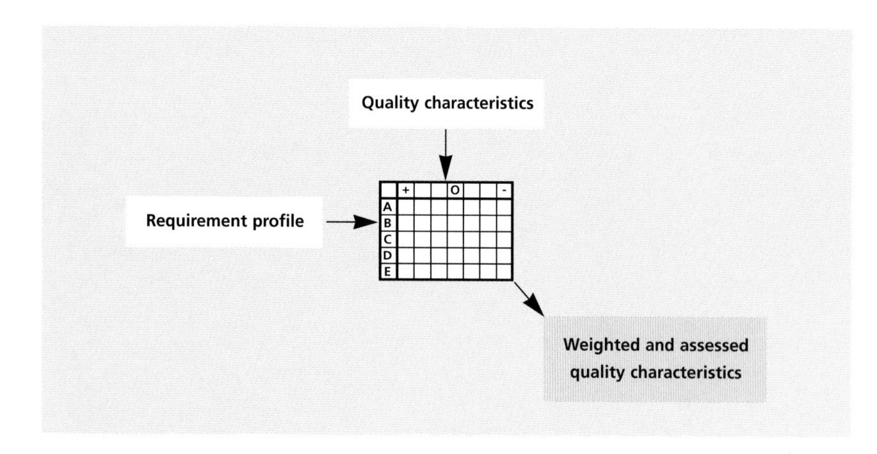

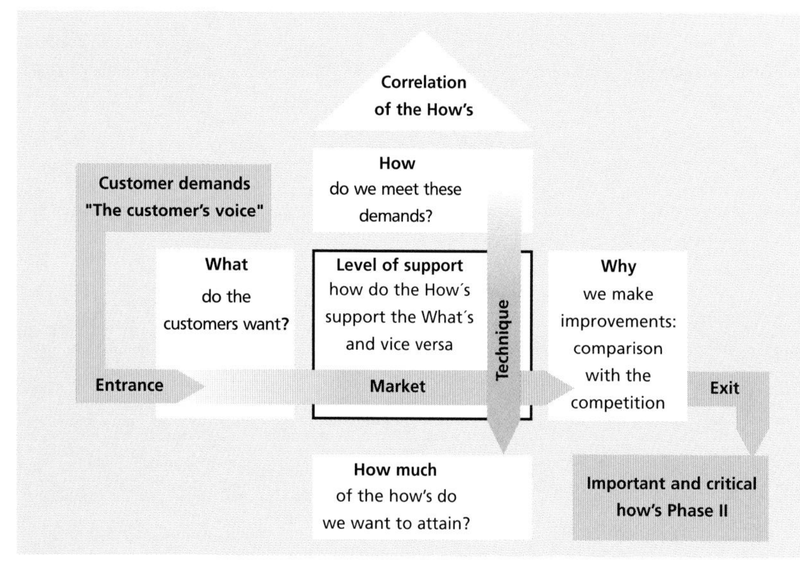

The House of Quality (HoQ)

The House of Quality (HoQ) helps to make the documentation of the thinking and planning results easily reconstructible. Its "rooms" correspond to ten steps that must be carried out in sequence. It starts at the entrance with the customer information from the marketing department and is then transformed into the language of the company in the second step.

Quality Function Deployment (QFD)

Quality Function Deployment (QFD) is a method for planning and developing quality functions corresponding to the quality characteristics demanded by the customer during each phase from research phase, through product development and production, and right up to marketing and sales.

II.) From Product Definition to Concept

The requirement profile lists the structured and weighted requirements of the service system. The requirements of the developing company are listed along with provider, operator, and user demands. The requirement profile may be compared with existing products to illustrate the level of performance.

The quality characteristics are summarized in a list that includes all the technical characteristics for assessment of the service system. The list includes data such as geometrical size, weight, force, velocity, precision, and also information about surface characteristics and consistency. For the individual quality characteristics, optimization procedures and target values are defined.

Quality Is the Top Priority

If the requirements and quality characteristics are arranged and compared in a House of Quality (HoQ) according to the Quality Function Deployment (QFD) method, a weighting of the quality characteristics with the help of the requirements can be achieved from the individual dependents. The weighted quality characteristics represent the assessment construction for further development. All functions, subsystems, and parts must be developed so that they optimally fulfill the quality characteristics.

With the weighting of the quality characteristics, the definition of the product is complete, which now includes a description and a visualization of the idea, a technical-economical analysis, and a requirement and quality profile.

品質 機能 展開

HIN SHITSU KI NO TEN KAI

Japanese version of QFD

品質	機能	展開
Quality	Function	Distribution
Characteristics	Mechanization	Diffusion
Attributes	Task	Development
Quality label		Evolution

Functional Analysis and Structuring

The aim of the functional consideration of service systems is to develop functional solutions for performing services, without using direct components or software modules as part of the solutions. In this way, the system can be regarded as a harmonious interaction of software, control, and mechanics.

The functional analysis usually begins with the system's core or key function. The functions become more and more detailed and are broken down into subfunctions until a sufficiently accurate description of the whole process can be shown. A function is generally described using an infinitive. For example, the key function of a climbing robot's intelligent vacuum gripper is "to attach" whereas its subfunctions explain that the gripper has "to create," "to maintain," and "to monitor" the vacuum inside its suction cup.

The functions and subfunctions must then be put into a structure that shows their connections. Various techniques are available to demonstrate the structure of the functions; most have tree or class-type structures. The choice of technique to be used is dependent primarily on the complexity of the service system and the weighting of software in relation to mechanics. The structuring of the functions should aim simultaneously at reorganizing them. By combining or substituting, the structure is constantly checked during its construction for optimization potentials.

The completed structure of the functions can then be reconverted into a structured list. This list of functions is then placed in a matrix with regard to the weighted quality characteristics according to the Quality Function Deployment method. A weighting of the individual functions can be ascertained from the interdependencies. Furthermore, the function that is responsible for fulfilling each quality characteristic can also be determined.

Finally, the functions go into more detail using process values containing the performance values of the functions. To some extent they can be derived from the comparison with the quality characteristics.

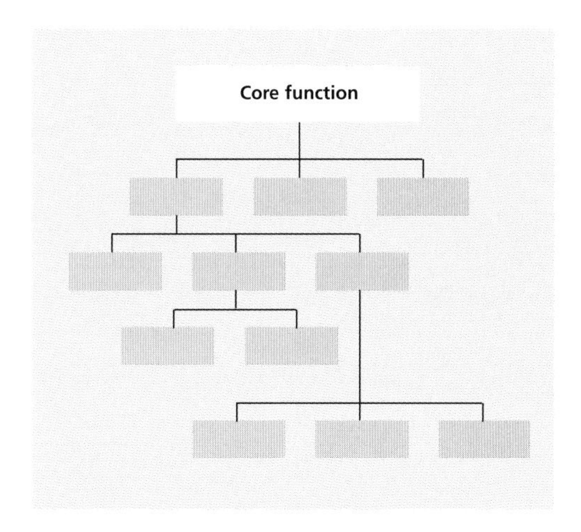

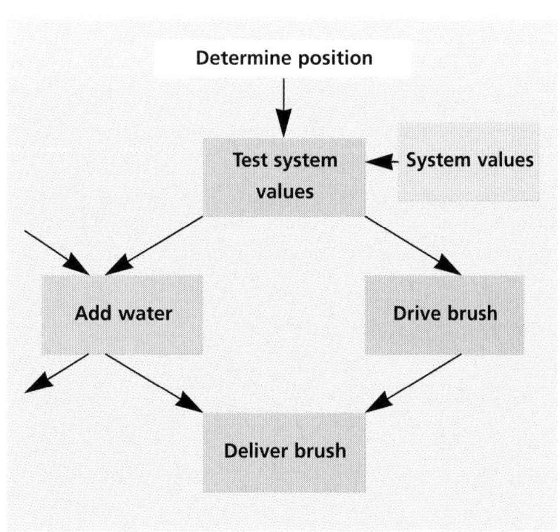

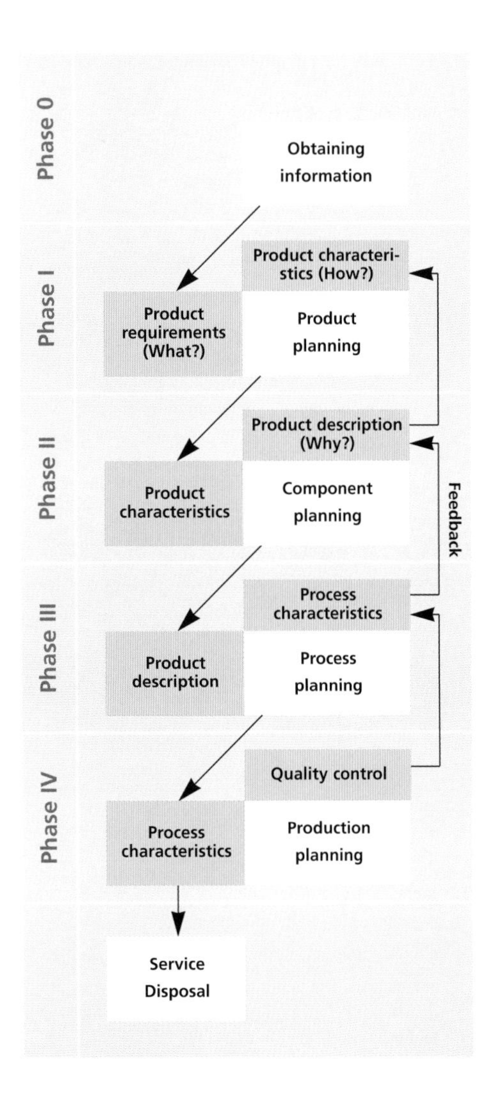

the 6-3-5 method
The 6-3-5 method is a team-based brainstorming strategy with which optimal problem solving can be achieved.

Finding the Concept and Assessment

Based on the functional structure, alternative technical solutions are developed to carry out the individual functions. The 6-3-5 method, for example, can be used to help generate ideas for solutions. The suggested solutions are collected together and pre-assessed with regard to their technical feasibility. The chosen solutions are then developed into concepts.

The alternative concepts are compared and assessed according to the Quality Function Deployment Method in matrices with regard to fulfilling functions and quality characteristics. The result is a selection of individual concepts that must then be united into a collective concept.

III.) From Concept to Product

A collective concept is created from the individual concepts, which at the same time characterizes the service system in its construction (subsystems and parts). Then the functions of the subsystems and parts must be classified. For this, the list of subsystems is compared with the list of functions, and the interdependencies are shown in a matrix. A further comparison enables the quality characteristics of the subsystems to be classified. In this way, quality characteristics can be derived for the individual subsystems and parts.

Simulations and Practical Tests Reveal Errors

For the arrangement of the subsystems, concepts are developed in the same way as for the functions. Simulation programs enable a functional and organizational assessment of the individual concepts. They are defined and converted into a plan that is then developed into technical drawings and software design. To check the individual functions and the collective function, functional models and prototypes can be constructed, which must then undergo real tests.

Adequate practical testing under real-world conditions is particularly necessary in the case of systems that interact directly with humans. Humans show an inestimable variety of rational and irrational reactions. The widespread fear of the "robot gone crazy" should be avoided by performing adequate functional tests so as to reduce the likelihood of a malfunction under general operating conditions. The safety of individuals in the surroundings of a service robot must be absolutely guaranteed in every situation.

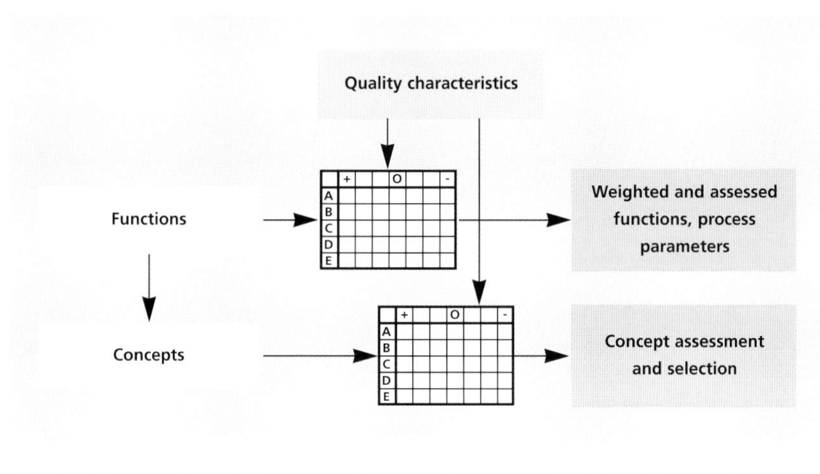

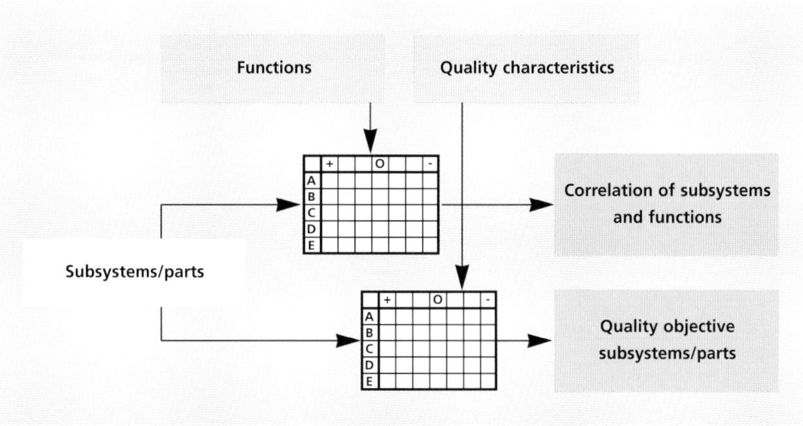

Basic Functions of a Service Robot

Orientation in Motion

There are three crucial questions for each freely mobile robot:

- Where am I?
- Where do I want to go?
- How do I get there?

It will always find the answer by using the same basic functions, independent of their application or the place of implementation.

A service robot usually works in unfamiliar, potentially changing surroundings. With the aid of artificial sensory organs called sensors, environmental perception supplies the necessary information about its implementation area. Based on this information, the service robot can form an internal picture of its surroundings, called an environment model. This model serves as the basis for determining the robot's position and orientation. The accompanying positioning function answers the question, Where am I? and may, for example, use distinctive points in the surroundings to accomplish this.

The answer to the question, Where do I want to go? is determined by the tasks to be performed by the service robot. A transportation task is defined by a starting point and a target point. In the case of a cleaning robot, the area to be cleaned is given instead of a defined target point.

Both a global motion planner and a localized path guidance system are needed to reach the target. The path guidance system guarantees that the task will be performed in optimal time. For this, tracking points based on the environment are supplied, which determine the service robot's path of travel. Using interpolation, the path planner calculates in detail the robot's path between these tracking points; should obstacles not originally included in the environment model bar the way, it makes evasive maneuvers possible.

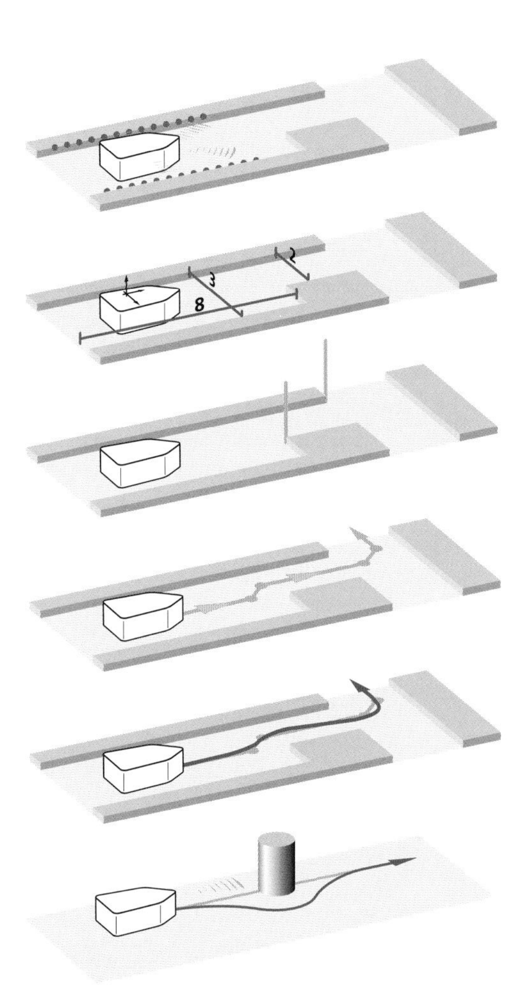

From top to bottom:
Environmental perception;
Environmental modeling;
Positioning; Path planner;
Path guidance; Avoiding obstacles;

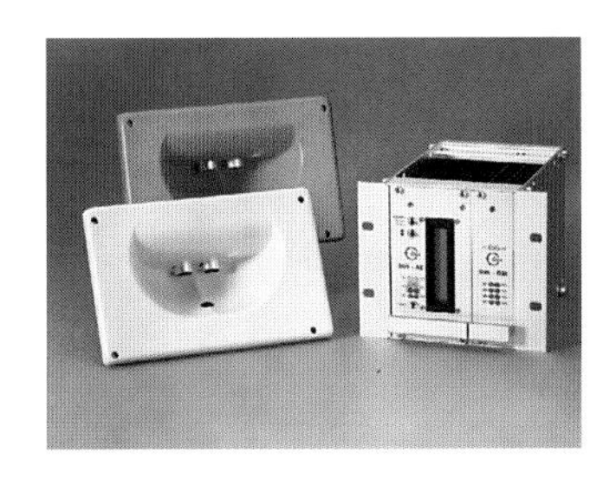

Ultrasound sensor with safety license, Elan Schaltelemente GmbH, Wettenberg, GER (left); 2-D laser scanner PLS, Sick AG, Waldkirch, GER (right)

Perceiving the Environment

A service robot's area of application is not usually laid out to suit automation in the same way as in manufacturing. A service robot is implemented in an unstructured, usually unknown, and altered environment in which people may be moving about.

Special demands are made on the sensory perception of the surroundings because of the lack of spatial division between the service robot's working area and the area where people and objects are found. The external sensory analysis, acting as an interface between the robot and its environment, is a basic technology in the utilization of service robots. Ultrasound sensors are well-suited because they are a simple and inexpensive way to determine distances from objects. Laser scanners and, increasingly, image-processing systems are also being implemented.

By moving an ultrasound sensor or using special sensors based on multiple arrangement, the direction of measurement can be altered. These ultrasound scanners supply an image of the surroundings similar to that of a radar map.

A distance measurement value is calculated from each angle of reflection. Due to the fact that the measurements from different ultrasound sensors influence each other, they are carried out frequently and in quick succession. The transit time of airborne sound limits the rate of measurement, however, which makes this method inadequate for fast-moving robots. For this reason, optical laser scanners are being used more frequently, as they are able to measure the exact position of an object in pico-seconds due to the fast transit time of laser light (10 -12 seconds). The laser beam is optically diverted, using a rotating mirror, for example.

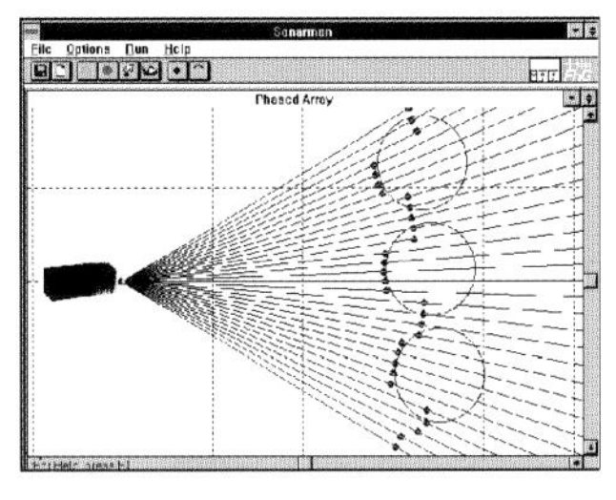

Ultrasound scan of standing paper rolls

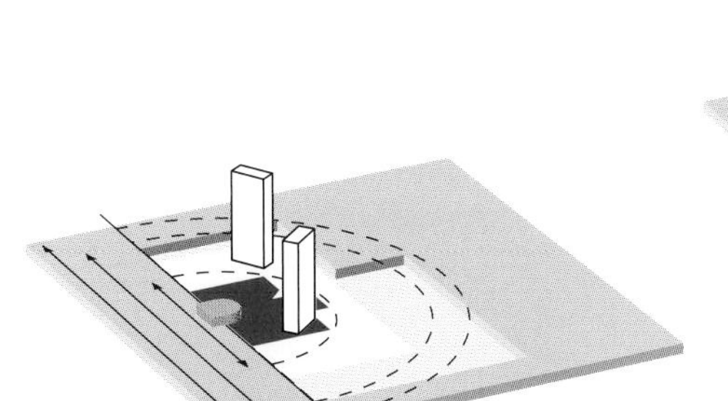

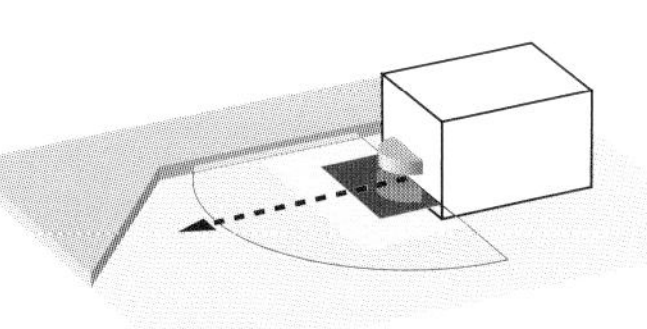

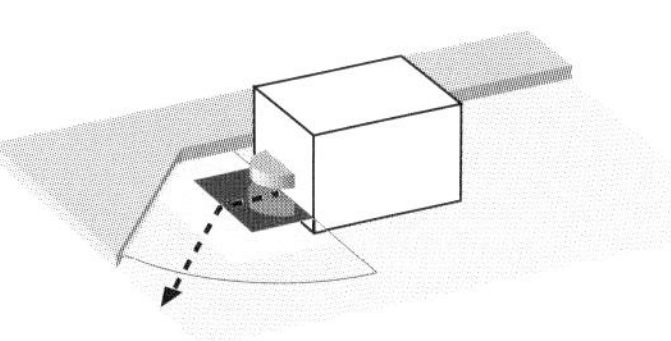

Spatial detection using optical sensors

Programmable warning field for reducing speed when approaching an object

Programmable protective field. If an object or person enters the protective field, an emergency shutdown is triggered.

Laser scanner

Environment Modeling

For safe, collision-free and target-oriented movement, a robot needs information about the actual status of its surroundings. To eliminate measurement errors, the measurement data from several sensors (ideally, sensors using varying physical principles) are calculated together and merged to form an environment model. Many current scientific papers are dedicated to the proper method of fusing the data from the multi-sensors.

In order to process all the information about the surroundings, various algorithms are used to reduce the volume of data. The first methods are based on a grid model of the environment. For this, a value is given to each point in the grid, indicating whether an object is to be found in the corresponding position. Because a grid model has a high memory requirement, methods have been developed to allow areas to be summarized together.

Using an iterative division of an area into four quadrants, the energy involved in the calculations can be reduced using the quadtree model. Also, edge models are often used where the boundary of the surroundings is recorded merely as a polygon shape.

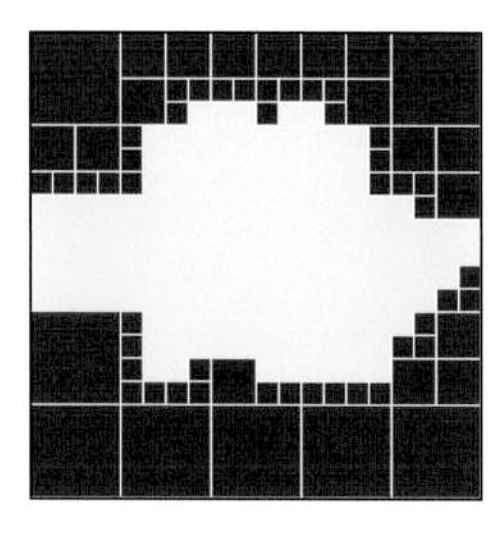

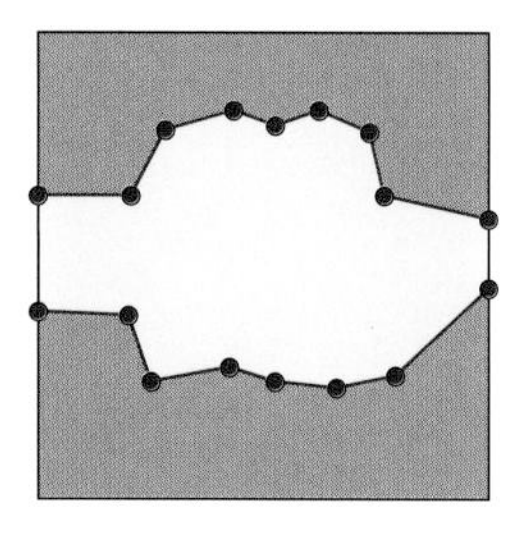

vacant
occupied
unknown

1. Grid model. Basic elements: quadrants or cubes of the same size

2. Quad/octree model. Basic elements: quadrants or cubes 2^n in size

3. Vector model. Basic elements: Lines or polygons

Positioning

Knowing the current position of the robot is essential for the path planner and the path guidance system. Distance traveled is determined in the same way as an odometer: by measuring the revolutions of the wheels. Because small errors constantly occur from influences such as slipping, the accuracy of the interlayer coupling is reduced as the distance traveled increases. Reference measurements ("supporting measurements") are, therefore, necessary at regular intervals to compensate for errors caused by slipping.

Independent of external references, the orientation is determined by sensor systems using gyroscope or revolution rates. These measuring systems have long been implemented for posi- space travel, and for military purposes, such as missiles and military vehicles.

The first gyroscopes used the physical principle of conservation of angular momentum. A rapidly rotating mass is suspended on gimbals in a frame. By conservation of angular momentum, the body retains its spatial position so that the torsion of the machine can be measured at the point of suspension of the gimbals. In the same way, there are structures with only one or two degrees of freedom. In other structures, the rotating velocity of the measuring system is calculated from the torque, the latter needing a constant position relative to the frame in order to maintain it.

In addition to these classical gyroscopes, new constructions have been developed, such as solid state gyros containing oscillating structures. These constructions are driven either by electrical fields or piezoelectric crystals. A rotation at the level of the structures produces a Coriolis force proportional to the rate of revolution, which in turn produces further oscillations. The phase displacement is measured and transformed into the sensor's output signal. In current research, these structures have been miniaturized and then integrated into the same chip as the analyzing electronics.

Laser and fiber gyroscopes give the highest level of precision, are strongest, and offer the widest range of measurement. Optical gyroscopes use the Stagnac effect, in which the displacement of the length of the light waves occurring when a ring-shaped path of the rays is turned, is analyzed. The output signal is directly proportional to the rate of revolution.

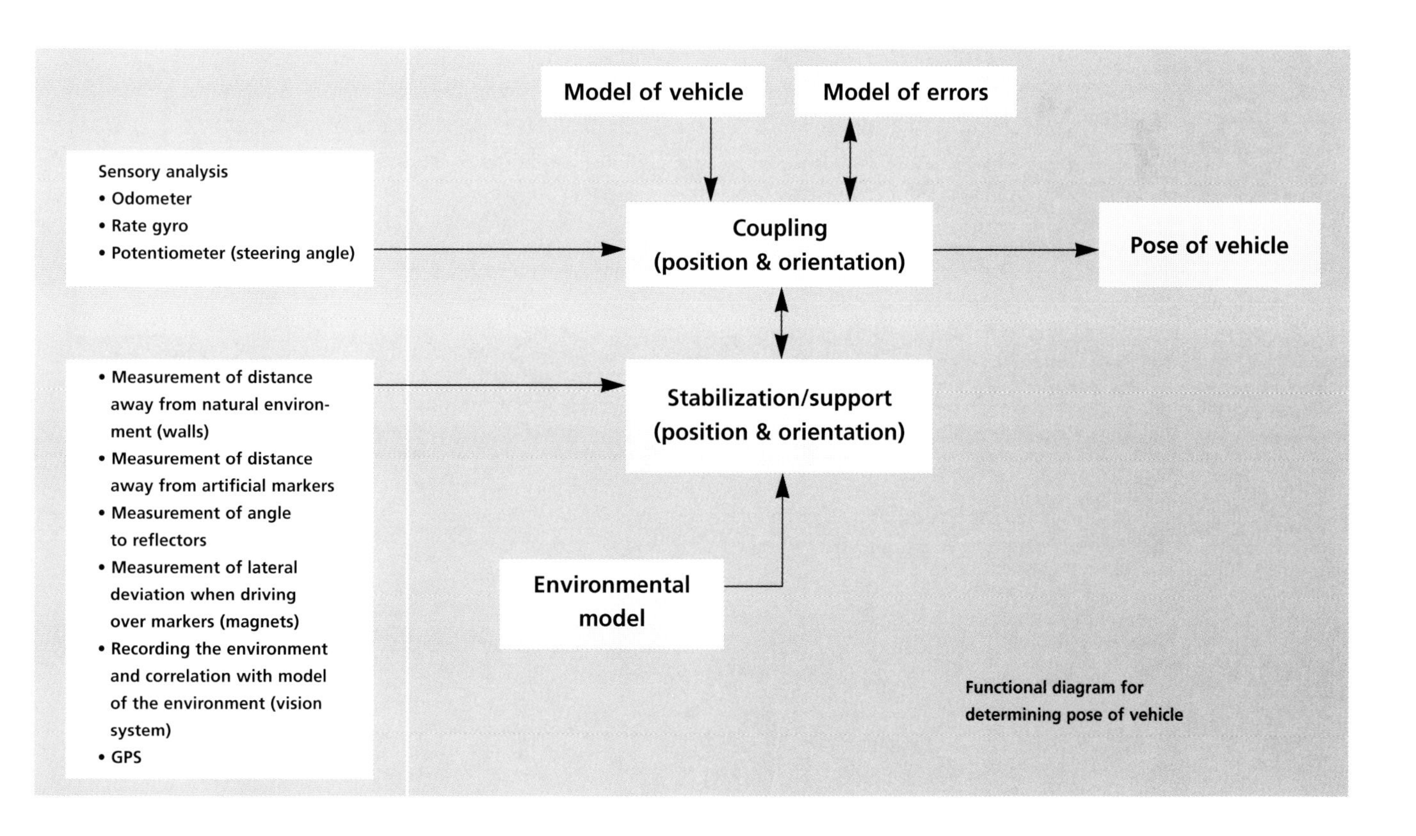

Functional diagram for determining pose of vehicle

The Path Planner

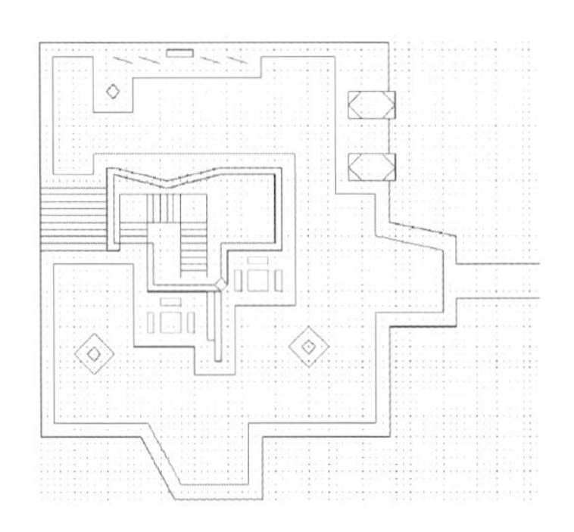

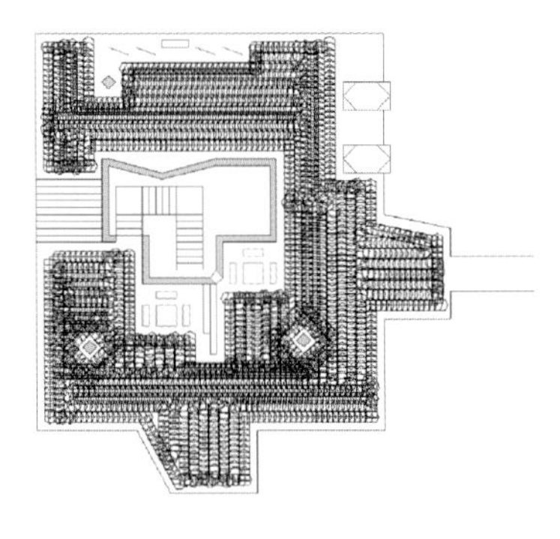

Devising a cleaning path based on a linear environment model

At the beginning of an operation, the path planner decides on a general course of movement, for example, in the form of intermediate positions. Depending on the service robot's job, various strategies are implemented.

For transportation tasks, intermediate positions, which are based on the environment model, are planned so that the target can be reached in the shortest time possible. The system maintains a safe distance from obstacles at all times. In the case of automatic floor cleaning, the intermediate target points must be placed in such a way that the entire area to be cleaned is covered. The service robot must be able to drive close to obstacles and to the boundaries of the area so that the uncleaned area is as small as possible. Depending on the dimensions of the room, algorithms may be required to determine the optimal cleaning path.

Path Guidance

The path guidance system is responsible for planning sensible paths of movement and for carrying out localized path corrections.
A sensor-based path guidance system should be able to avoid collisions and carry out evasive maneuvers. Numerous algorithms have been developed which, based on information obtained from the sensors, guide the service robot around an obstacle, allowing it to avoid collisions.

With the potential field method, a virtual repulsive force is calculated from the distance away from obstacles. The total force, calculated from the sum of all the obstacles, determines the direction in which the service robot will move. This method can be described in terms of ascent and descent, with obstacles representing the summits of individual mountains, and the target position representing a valley. The service robot's behavior corresponds to that of a ball rolling through this landscape into the valley.

Weaknesses associated with this method, for example, the tendency to oscillate in bends in the path, have brought about the development of improved strategies; one example is the vector histogram, in which the density of obstacles in various directions is considered when determining the direction of travel.

Safety and Collision Protection

When robots are used alongside of people, the safety of the system is of particular importance. In addition to implementing obstacle-avoidance strategies, a redundant system working on a much lower functional level is integrated in order to fulfill these safety requirements. This system recognizes imminent collisions and prevents them from occurring by stopping the robot in time. To do this, tactile sensors which also function as buffers, bumpers, and contactless ultrasound sensors and laser scanners are used.

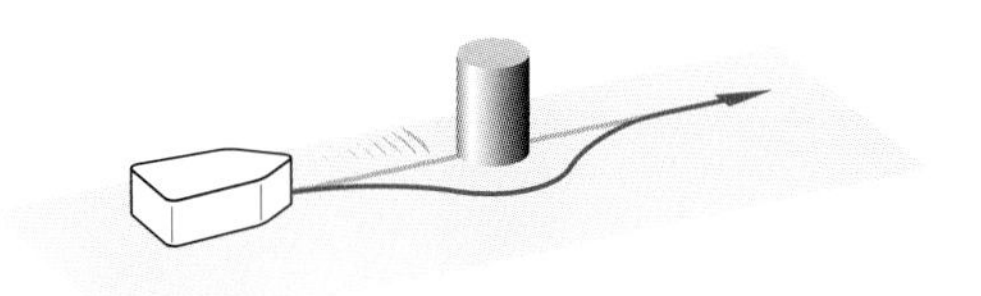

Avoiding obstacles

Ultrasound

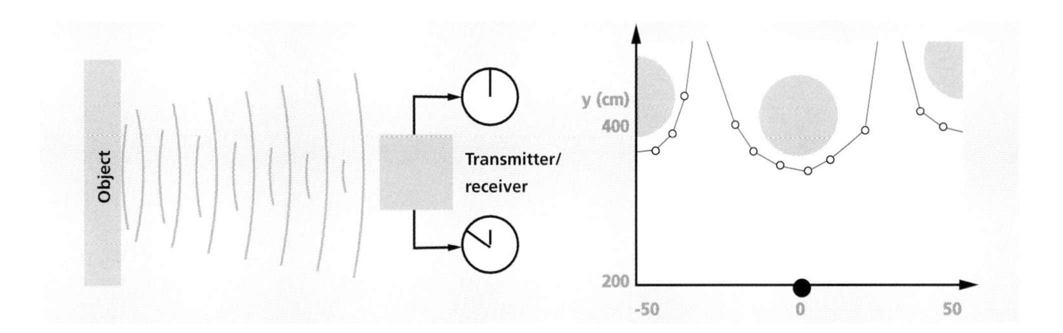

Laser scanning

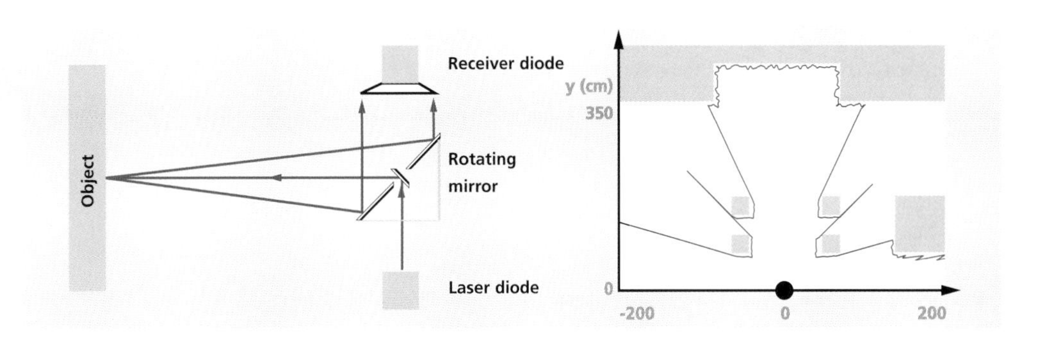

CCD camera

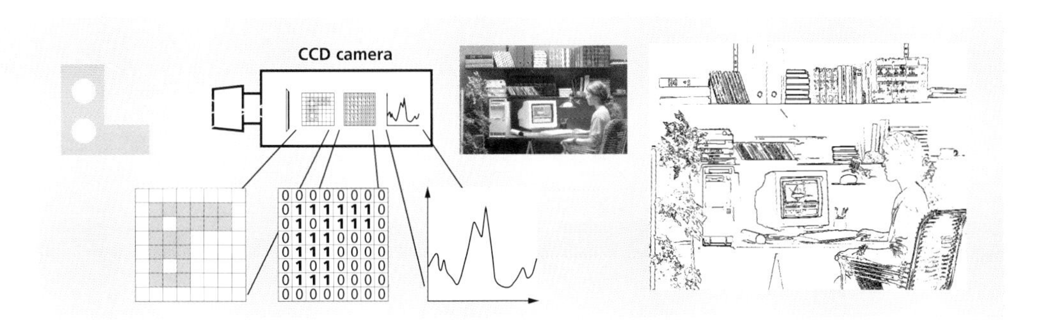

Scenarios - Service Robots in Action

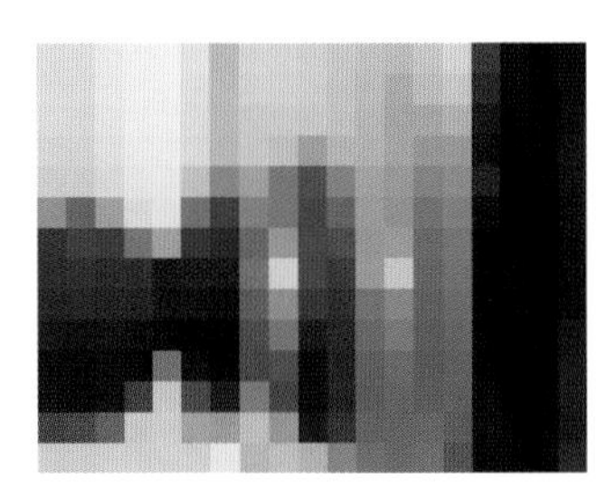

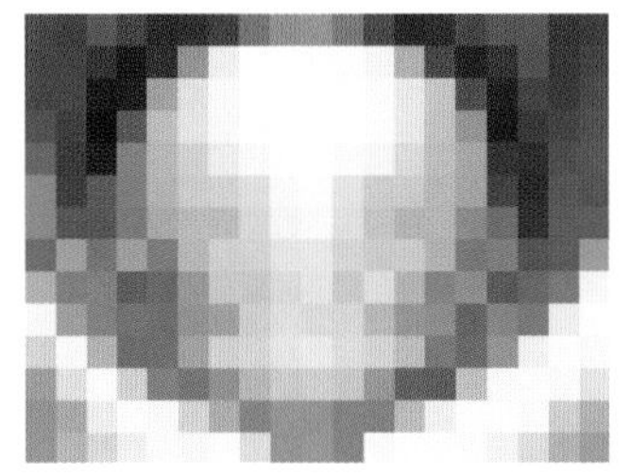

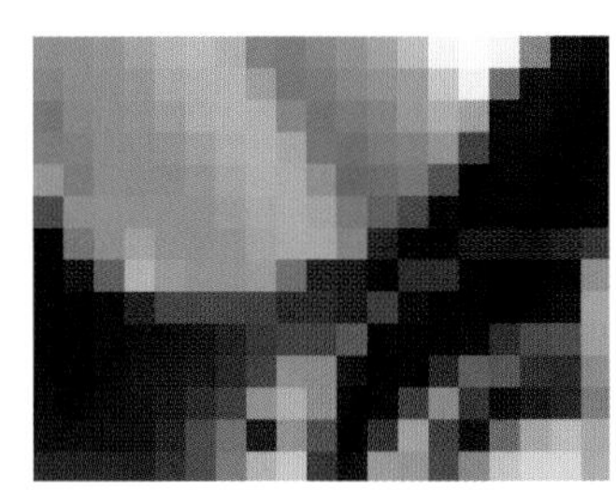

Refueling

1920

Gas Stations Made Simpler

Nearly everything having to do with automobiles is automated nowadays. Only refueling still takes place as in the days of Carl Benz and Rudolf Diesel: by hand. The air at a gas station is heavily laden with gasoline and benzene vapors, and is hazardous to everyone's health.

At the end of the nineteenth century, the number of motor-driven carriages was still so small that an entire gas station could fit onto a horse-driven wagon and be pulled from one town to another. The first stationary gas stations triggered free competition. The gasoline pumps belonging to the bigger companies stood peacefully next to one another and the customer was able to choose his favorite grade on the spot. Brand-name gas stations developed with the increasing number of vehicles and types of fuel.

Over time, individual subsystems of the gas station, such as the gasoline nozzle or the pump station, have been constantly improved. However, engineers have largely overlooked the potential of optimizing the refueling procedure itself, the insertion of the nozzle into the gas tank.

There Are 50,000 Gas Stations in Europe

A considerable potential benefit exists here. In Germany alone, there are approximately 10,000 gas stations. In the whole of Europe there are 50,000. Average refueling in Germany (35 liters), carried out manually and including payment time, takes seven minutes. A refueling robot can reduce this time to just two minutes.

A more important reason for using refueling robots is the ecological aspect. In Germany, refueling produces over 10,000 tons of toxic fumes annually. Over 90% of these emissions, which are dangerous to both the environment and human health, could be eliminated if refueling were carried out by a robot. Present day nozzles manage just 60% of these fuel vapors. Eliminating the need to exit the car to refuel is an important advantage, particularly for the physically challenged. Refueling by robot achieves a higher level of safety and comfort for the automobile driver.

1927

1950

1970

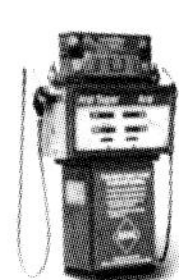

1980

1990

Fraunhofer IPA refueling robot - tomorrow's gas station?

Yesterday's gas stations

Finding the Gas Tank

With the current variety of private cars, the question Where is the tank? is not so easily answered; without retrofitting new technical parts, automatic refueling is not possible. The situation is different, however, in the case of fleets of commercial or publicly-used vehicles, such as public service buses. Here, the driver is less inclined to refuel his vehicle himself. Manual refueling wastes a huge amount of time when carried out several hundred times a day.

In the 1980s, as industrial robots started their triumphal procession and automation found greater popularity outside of factory walls, the utilization of robots for refueling vehicles was also considered. In this area of application, five different robot-operated refueling systems have been developed to date, and some of them are already being used.

Automated Refueling of Alternative Fuels

A distinction is made here between systems suitable for fossil fuels and systems suitable for alternative fuels, such as hydrogen. When the conversion from fossil to regenerative fuels takes place, manual refueling will no longer be possible for safety reasons; people will want their refueling with hydrogen to be carried out only by robot. A secure and absolutely air-tight connection between the vehicle and the pump nozzle is essential.

Energy requirements in the third millennium

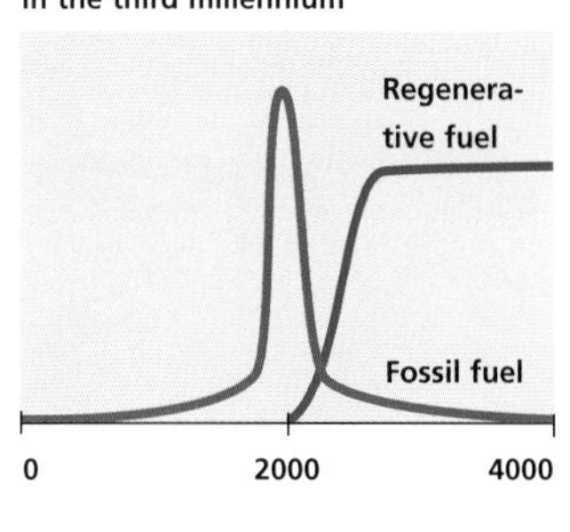

Fleets of Buses Can Be Refueled Quickly

Robin, the robot-controlled refueling system, with its individual network docking procedure, has been refueling the state-run buses in Saarbrücken, Germany, since 1993. The Saartal Linien, the local transport companies in Saarbrücken, were the first German transport operators to operate a fully automated refueling system. Robin was developed and manufactured by the Anton Bauer GmbH in Dillingen, together with the Raab Karcher Tankstellentechnik GmbH in Hamburg, Germany. After more than two years of development, the robot-assisted refueling system was put into operation in 1993 for 150 public service buses. The high setup costs incurred by the drivers when refueling the buses was the reason for developing the pilot system. With a setup time of five minutes, not including the actual refueling procedure, the Saartal public service buses, being refueled about 140 times a day, had annual setup costs of over $90,000 at the time.

Bus-refueling system Robin

Accurately Positioning the Nozzle into the Filler Opening

When the bus is driven into the refueling lane, a transmitter in the front section of the roof of the vehicle sends all the necessary data to the refueling robot. Meanwhile, the driver steers the bus into a hollow made for the wheel; this positions the front right wheel of the bus and marks a reference point for the robotic system, from where the distance from the filler opening is known.

The robot then moves along rails at the height of the filler opening, parallel to the longitudinal axis of the vehicle. The front jib holding the gas nozzle is moved pneumatically. Precision adjustment takes place by way of five inductive distance sensors, which measure the path to the contact plate at the filler opening. When the gas nozzle has been inserted, the filling process commences. On completion of refueling, the robot moves back to its initial position and the bus may leave the refueling lane.

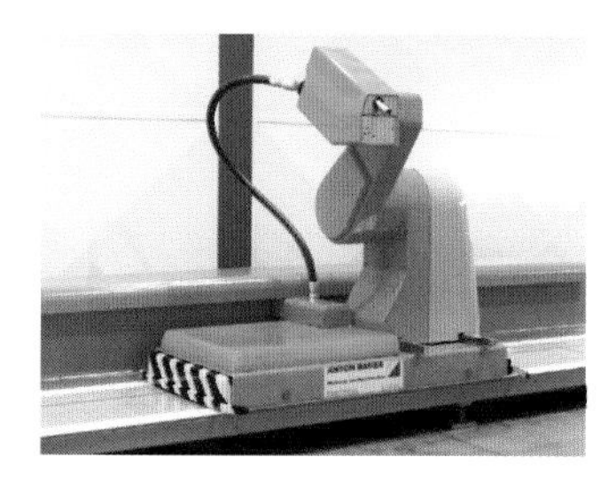

Bus-refueling system Robin, Anton Bauer GmbH, Germany

Automatic Forecourt Staff at Conventional Gas Pumps

Another refueling system, Oscar, is much more similar to existing gas pumps than Robin. The first Oscar system was built by the French company Robosoft at the end of the 1980s and has been further developed up to the present day; the most recent system is Oscar MK5. Three systems are grouped around the robot for data input, vehicle identification, and refueling. A transponder attached to the vehicle provides vehicle identification information to a reading device at the gas station. The arm of the robot is vertically adjustable, and holds a slightly modified nozzle containing distance sensors. It is connected to a conventional gas pump via a hose.

Otherwise, Oscar functions in a similar way to Robin. The bus drives up to a mark in the refueling lane where a form of traffic light tells the driver to stop. The vehicle data is then read from the transponder; the vehicle and its dimensions are recognized. The robot positions itself approximately in front of the filler opening. The refueling system's sensors assist with the precision adjustment of the gas nozzle attached to the robotic arm and refueling commences.

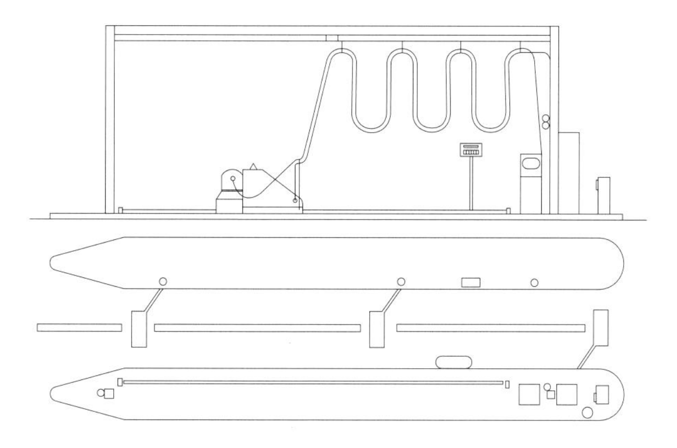

Bus-refueling system Oscar MKS, Robosoft, France

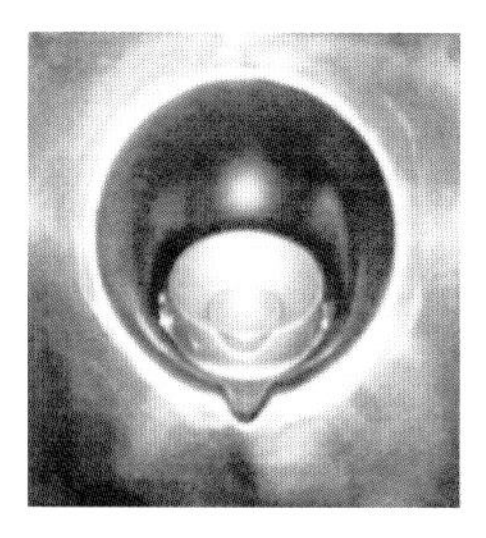

Filler opening

Bus-refueling system Oscar

The Robot Finds Almost Every Tank

Even the Smart Pump has to know where the filler flap is. Nearly every vehicle can be refueled automatically with this system if it is retrofitted with an active transponder and a filler cap suitable for handling by robots. The transponder transmits all the necessary dimensional data about the vehicle to the Smart Pump. Together with Shell Oil (USA), the Canadian company ISE Ltd. developed this automatic refueling system, which went into operation as a prototype in Sacramento, California at the beginning of 1997.

The Smart Pump robot is a classic portal construction with two linear axes carrying the cross arm. The telescopic axis, affixed with the refueling head, is attached to it. The receiver for the data from the vehicle transponder is at the portal, as well as a video camera for locating the contours of the vehicle and a safety sensory analysis system.

When the vehicle's engine is switched off, the operating terminal glides up to the driver's window and the refueling request is entered via a touch screen. The robot moves up to the filler flap, opens it, and manipulates the filler cap which swings open inwards. The tank is then filled.

Smart Pump, International Submarine Engineering Ltd., Canada, and Shell Oil Products Company, USA

Nothing Can Work Without Cars Being Retrofitted

The system developed by the Swedish company Autofill refuels cars with lateral filler caps. It consists of a pump, a robot built on a linear axis, and an operating terminal connected to the gas station's central computer. Vehicles wishing to use this gas station must already have been retrofitted with a special filler cap.

If the driver stops the car next to the operator terminal in such a way that he is able to use the entering device without having to leave the car, the car has automatically been correctly positioned for the Autofill system. A retrofitted transponder, attached to the filler flap, transmits dimensional data to the system. The customer inserts his credit card into the vending machine and selects the amount he wishes to spend on fuel. Once the card's PIN code has been checked and the card authorized, Autofill begins the refueling procedure.

First, the robot positions its three linear axes so that the gas nozzle is placed in front of the filler flap. A vacuum gripper then opens the flap. The nozzle is positioned accurately with the aid of distance sensors and then inserted. Once the filling procedure is complete, the robot moves back into its initial position. Since December 1996, an Autofill refueling robot has been working at the OK service station in Mördby near Stockholm, Sweden, and another in Oslo, Norway, at the Norsk Fina gas station. Another system went into operation in Spring 1998 in Munich, Germany.

Refueling procedure, Autofill, Sweden

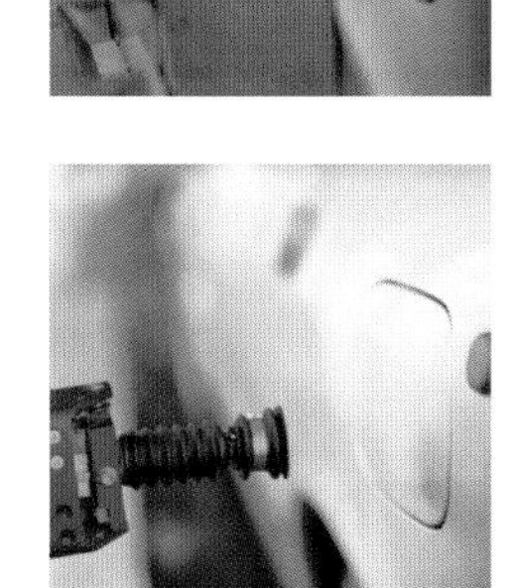

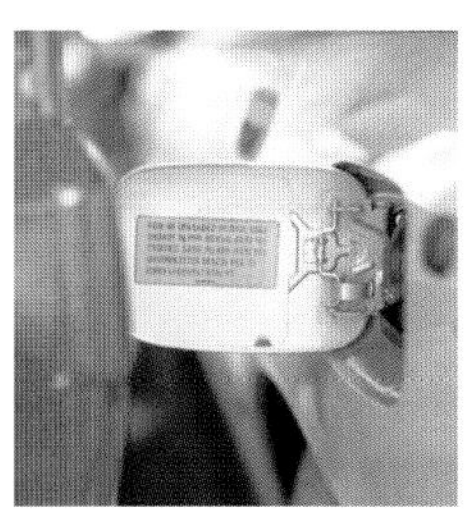

Refueling procedure, Autofill, Sweden

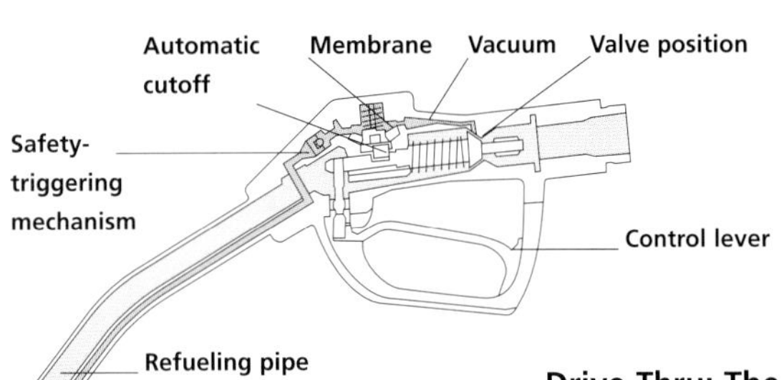

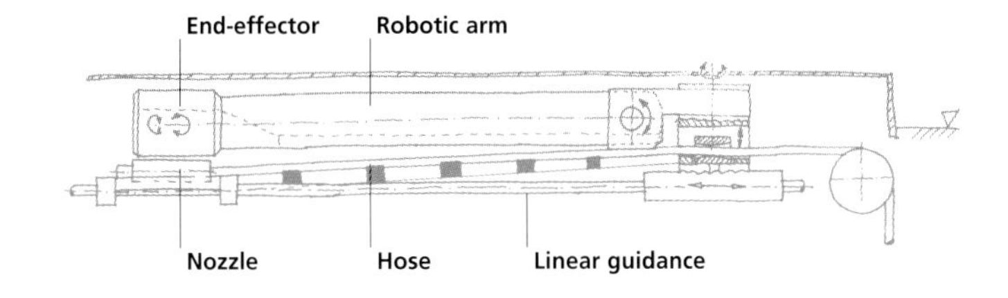

Comparison of manual and robot-operated nozzles

Drive-Thru: The German Vision of Refueling by Robot

At the end of the 1980s, an idea began to emerge in the heads of research engineers working for Aral, BMW, and Mercedes-Benz: fully automated, robot-assisted refueling, without harmful emissions, carried out in the shortest time possible and with the driver remaining in his or her vehicle.

Once the first concrete beginnings had been developed by the partners, the innovation management and conception were entrusted to the engineers of Fraunhofer IPA (Stuttgart). The expectations were clear:

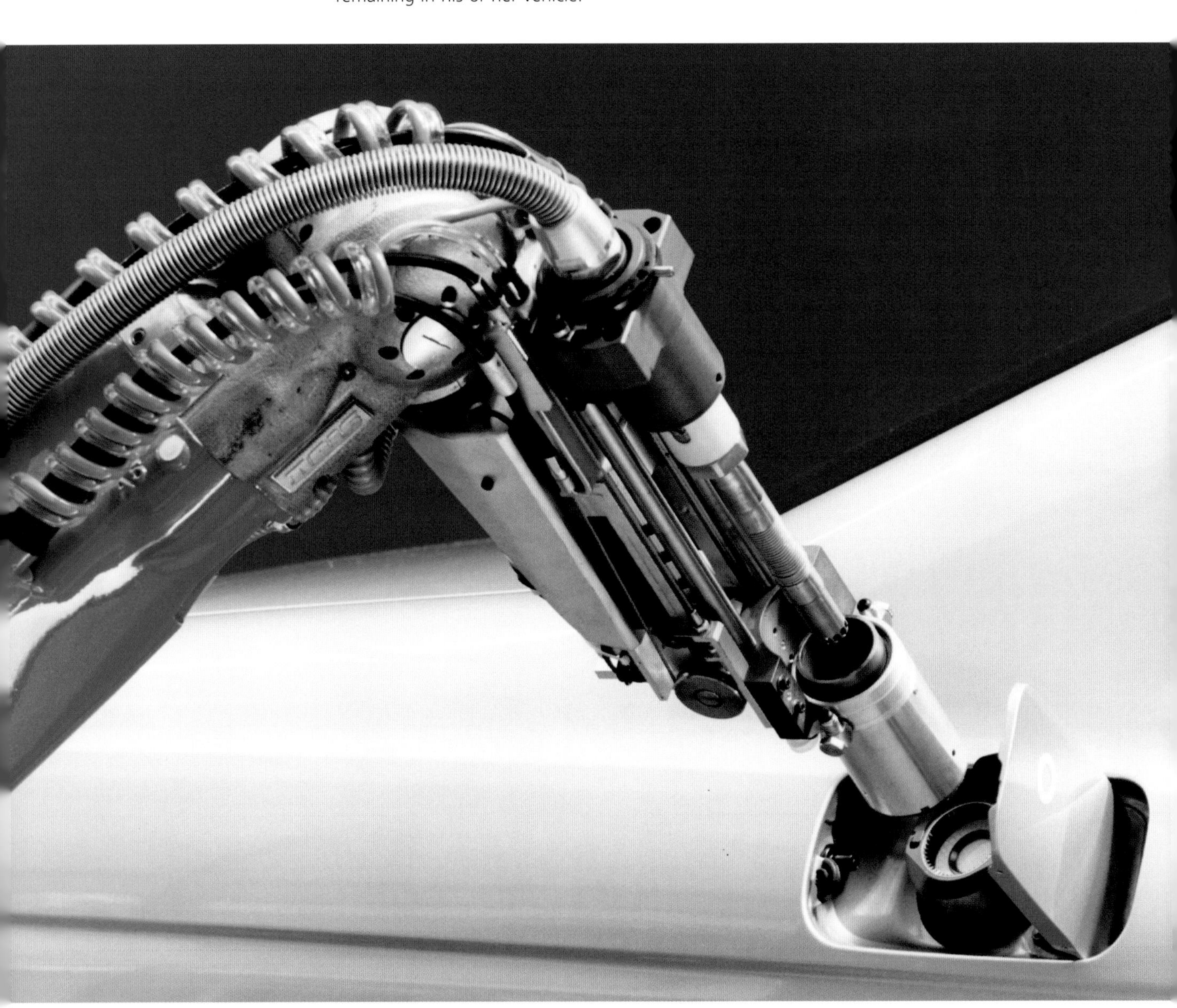

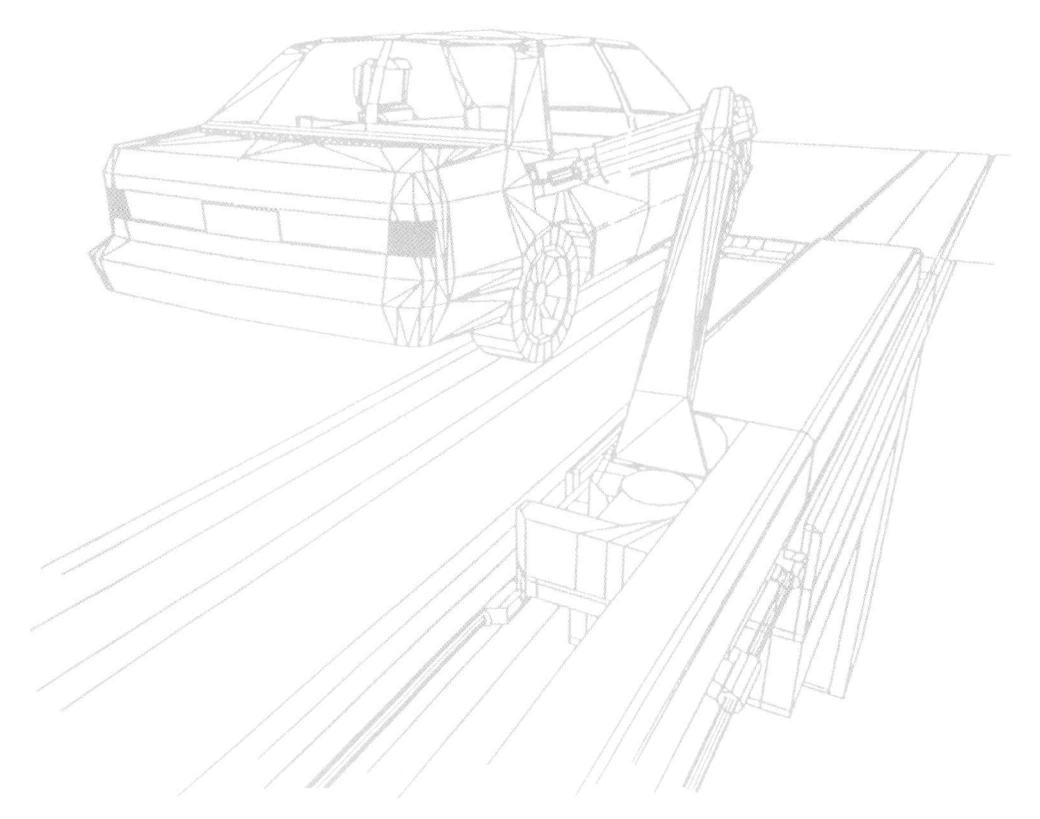

- Fully automatic refueling of a vehicle in two minutes. Over 80% of all vehicles with filler caps at the rear right hand side must be able to be filled up by a refueling robot.
- Minimal vehicle retrofitting costs.
- A selection of up to five types of fuel that would be delivered free of emissions and odors.
- Attractive yet functional layout of the gas station.
- Controlled and safe behavior of the system during unexpected occurrances, such as vehicular movement or the presence of people in the field of danger.
- The gas pumps must comply with refueling operation regulations regarding explosion hazards.
- The refueling robot should be able to be used economically as a series device.

Along the path from idea to prototype, numerous unconventional and innovative solutions were conceived and developed, and the refueling robotic system points the way for future applications of service robots.

Filling up a car using the Fraunhofer IPA refueling robot

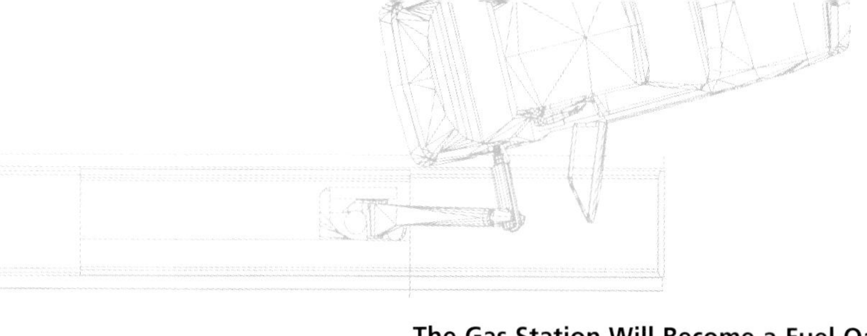

The Gas Station Will Become a Fuel Oasis

- Upon driving into the gas station, drivers will notice the absence of gas pumps; in its parked position beneath the pump island, the robot is out of sight. An extensive terminal forms the man-machine interface (MMI).
- Even before the customer has inserted his credit card, his vehicle will already have been identified. A passive transponder on the underside of the automobile transmits the vehicle type, choice of fuel, maximum amount of fuel to be delivered, and dimensional data about the vehicle on radioed request.
- Once laser scanners and CCD-cameras in both of the approach pillars determine the exact position of the vehicle, the robot's motion program may be generated. The robot leaves its parked position, carefully opens the filler flap, and establishes a firm and emission-free connection using the newly developed filling cap, which may also be operated manually. The customer selects the amount he wishes to spend on fuel. The actual filling procedure then begins.
- If the robot's environment should change during the filling procedure, such as in the case of an uncontrolled movement of the vehicle, the system always reacts in such a way that personal safety is guaranteed and no damage can be caused to the vehicle. The customer is able to leave the gas station after just two minutes. The robot then glides back into its hidden parked position.

transponder

"Transponder" is a merging of the words from "transmitter" and "responder". It denotes a passive transmitting system: the transponder only receives the energy to transmit coded data in a high frequency excitation field. Transponders have long since proved their worth in the prevention of shoplifting: They are hidden in price labels and set off an alarm at the exit barrier if they have not been deactivated at the cashier beforehand. The field strength of transponders is completely harmless.

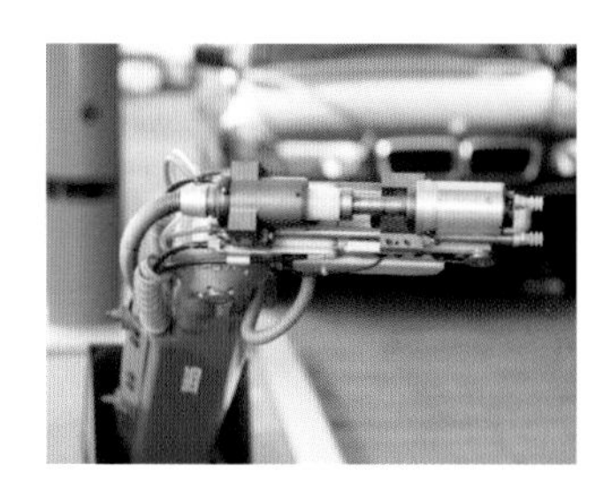

Automatic data recording

Selection of amount of fuel desired and paying procedure; refueling robot leaves its parked position

Automatic filling procedure commences

Completion of the filling procedure

Refueling robot glides back into its parked position

1985	1987	1990	1995	after 1997
First concrete idea	First talks between Aral AG, BMW AG and Mercedes-Benz AG	Syndicate contract and technology partnership, execution of the project begins	Presentation of the research project	Consortium opened for all interested parties; continuous study group; worldwide standardization being planned

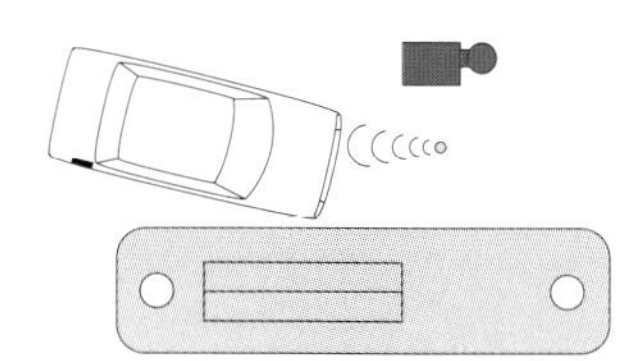

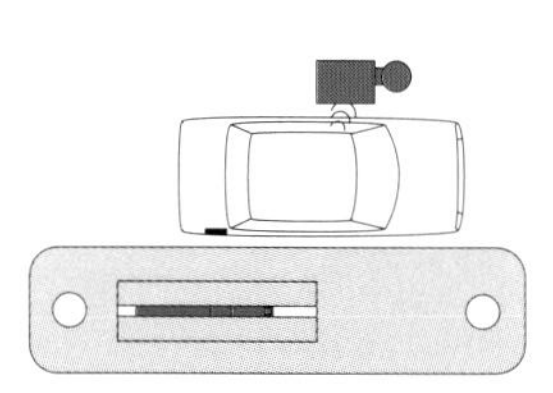

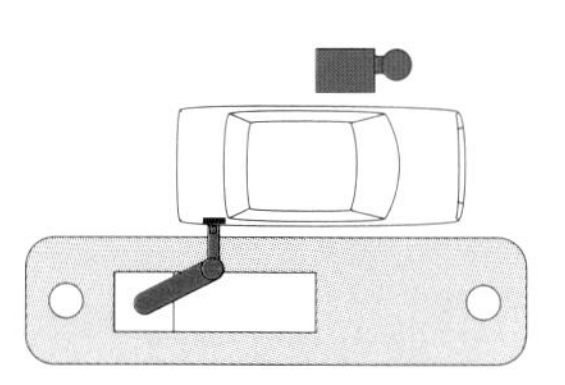

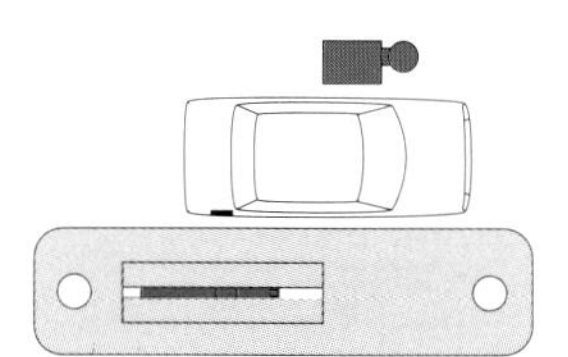

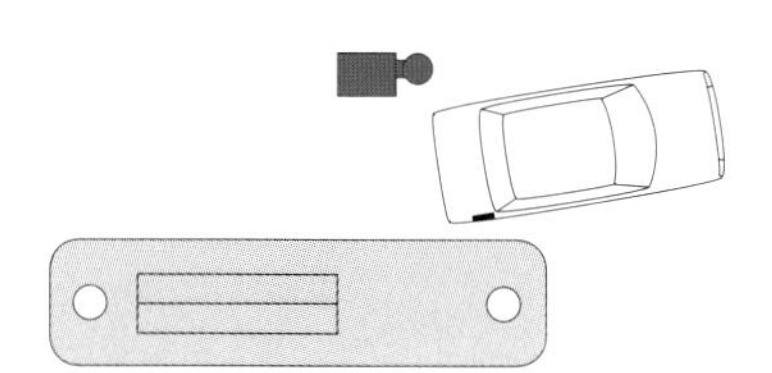

On the grounds of its institute in Stuttgart, Germany, Fraunhofer IPA has the first robot-operated gas station as a pilot system on display for demonstration purposes. A further pilot system for refueling vehicles with liquid hydrogen will soon be ready at the Franz Josef Strauss Airport in Munich: in a large-scale test, apron service vehicles are to be operated using liquid hydrogen.

A Commitment to the Future

The market potential for automated refueling systems is huge. A conventional gas pump with five hoses costs around $44,000, and the costs of a robot-operated refueling system run to less than $83,000. It requires less space and handles a considerably higher throughput of vehicles.

Environmental aspects, as far as conventional fuels are concerned, and safety aspects in the case of alternative fuels, will sooner or later make the manually operated gas station obsolete. So why wait?

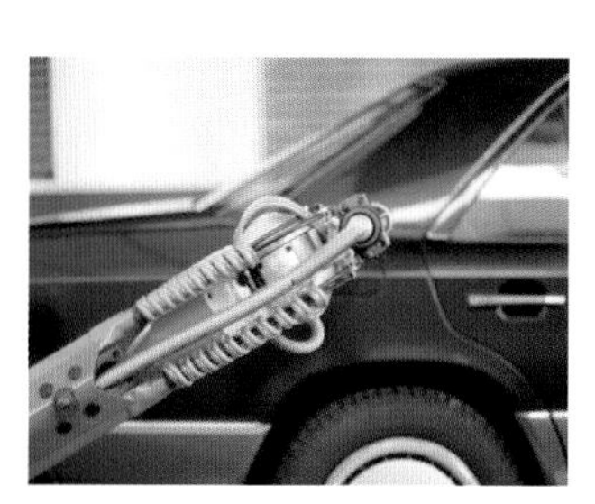

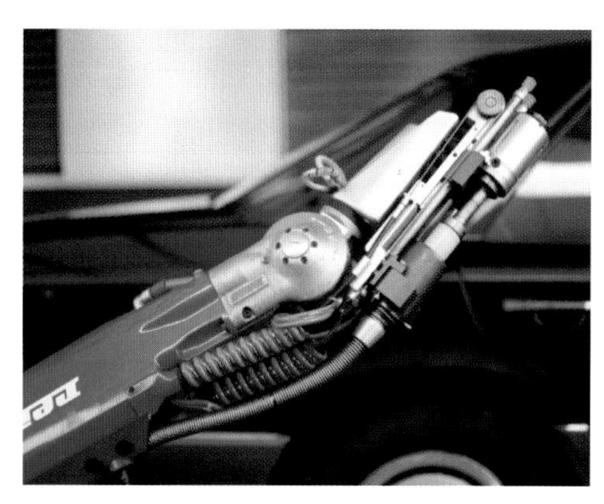

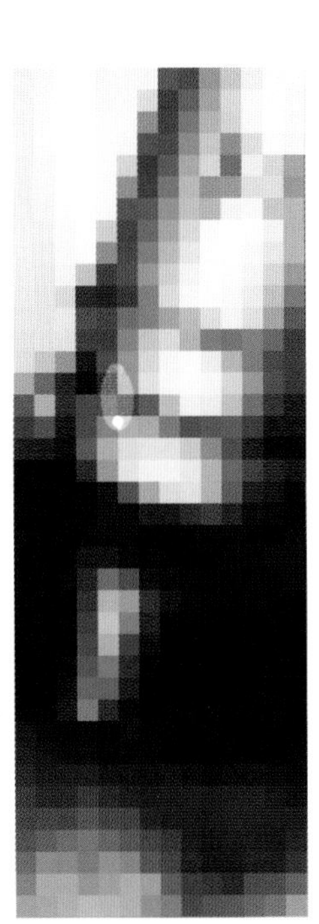

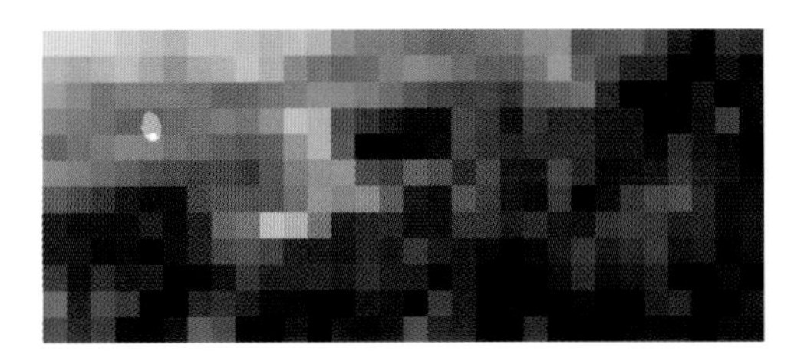

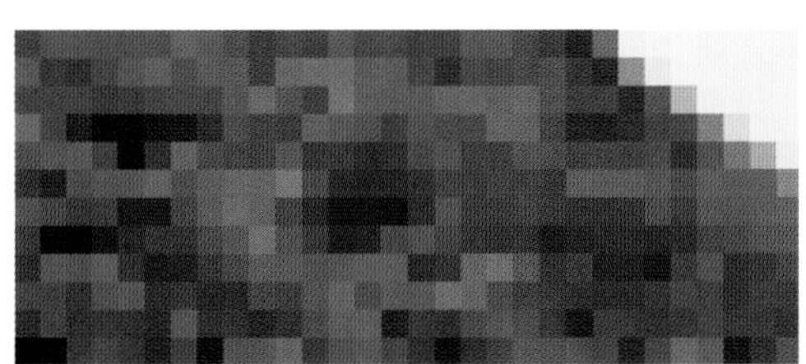

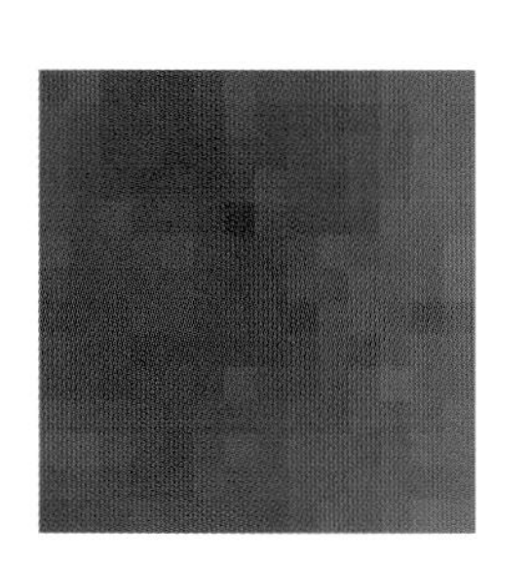

Agriculture and Forestry

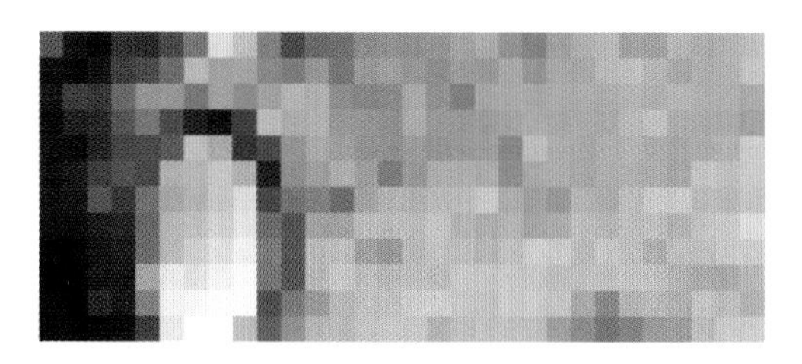

Progress for Nature

Today, if a farmer or forester wants to remain competitive, he has to implement innovative machines. Developments worldwide are giving man and nature a helping hand in many ways. For example, Japan's automatic spraying robot helps the environment by delivering the exact quantity of pesticide required. Also, robots are able to climb prickly palm trees and harvest fruit which, so far, man has not been able to reach.

Environmental considerations are playing an increasingly important role in forestry. Conventional forestry vehicles with wheels or with track drive belong on paths and tracks. If they drive over the forest floor, they can cause considerable damage to the land.

Walking forestry manipulator, Plustech Oy, Finland

Six-Legged Assistant

The Finnish company Plustech Oy has taken nature as its model and has developed the first walking forestry manipulator in the world. Their work opened up new possibilities for forestry unimaginable with conventional technology. Strictly speaking, the Plustech platform is not a service robot, but rather a sort of automated manipulator; nevertheless, the ingenious walking kinematics mechanism deserves recognition for its potential applications.

The partially automated six-legged manipulator can move safely over almost any terrain. The computer-controlled walking mechanism ensures that the weight of the device is always evenly distributed on the forest floor. The machine can be adjusted to suit varying ground conditions by simply changing its "shoe size", thus preventing soil erosion. The walking manipulator even feels at home on steep mountain slopes. Its maneuverability is excellent: It can turn on a dime without harming the ground. Tree roots are also subjected to less weight upon them.

The operator controls the forest manipulator using a joystick that simultaneously controls the speed of the vehicle. The manipulator arm integrated into the mobile platform is used to handle and transport tree trunks.

Tree-climbing robot, Cranfield University Centre for Precision Farming, Great Britain

Gently Picking Citrus Fruits

About ten different countries worldwide are experimenting with the development of picking robots. France, Israel, Italy, Japan, Spain, and the USA are in the lead. To date, robots for harvesting apples and citrus fruits are the most advanced.

From a merger of European research institutes and industrial enterprises and with the help of Professor Schillaci of Universita degli Studi di Catania in Sicily, one of the high-performance examples of this class of robot has been built as a prototype. The Italian company A.I.D. SpA and the Sicilian University have constructed a three-finger gripper with force momentum sensory analysis, able to grip a fruit securely without crushing it. Eight of these grippers, mounted on manipulator arms, are currently integrated into a track-laying vehicle.

Three-finger gripper, Universita degli Studi di Catania and A.I.D. SpA, Italy

An advanced image processing system locates the fruit by its color in the video picture of a color camera attached to the manipulator arm. Once the coordinates of the fruit have been transmitted, the gripper closes over it and a small blade cuts the stem. Up to 2,400 fruits per hour can be harvested using this type of construction.

By the time the prototype is ready for widespread use, the harvesting performance will probably have increased to 5,000 fruits per hour. The fruit will no longer be cut off the tree but will instead be twisted off by a turning motion of the gripper. Using this method, picking time can be reduced to approximately five seconds per fruit.

A similar robotic system is being used in France to pick apples using a suction gripper. Japan, Korea, and Hungary are also working on picking robots, but developments there are not so advanced. Current picking times of 16 seconds per fruit have no economic advantages for the user.

Ape-Like Tree Climbers

To date, some types of fruits have not been harvested at all. The South American macauba palm, with a prickly trunk 20 meters high, makes manual harvesting practically impossible. However, the surface yield of this tree is twenty times that of soybean plants. Agricultural experts in Brazil estimate that this plant could yield 10 million tons of oil annually and thus meet 20% of the world requirements.

This has prompted Peter Brenner, a construction engineer living in Rheinbach, Germany, to design a tree-climbing robot modeled on apes. The free-climber is made up of four flexed arms that clamp around the tree trunk. On climbing, the upper pair of arms is detached and a pincerlike mechanism pushes this pair of arms upward.

Once the artificial ape has clamped itself securely with both upper arms, its lower arms can then follow suit, assisted by the pincerlike mechanism. The drive mechanism consists of spindles with electric motors. Once it has climbed 200 meters in height, the onboard batteries are exhausted and must be replaced, preferably on the ground.

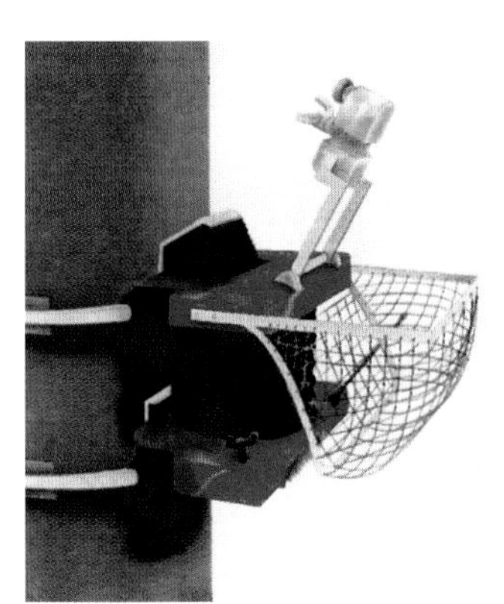

Tree-climbing robot
Peter Brenner,
Germany

Real Apes Are Stronger

The climber is directed from the ground using radio control. With approximately 60 watts of power, the robot, which weighs 8 kilograms, can achieve speeds of 0.5 meters per second. Tree trunks with a diameter of 15 to 40 centimeters are theoretically climbable. The autonomous robot should even be able to swing onto boughs, provided they do not branch off at an angle greater than 90°.

In the spring of 1998, the first tests of the robot ape were carried out. The path towards the finished product is particularly complex, however, due to the fact that various components suitable for use in the tropics are either not available on the market or are simply too heavy for practical use.

At Cranfield University in Britain, the Centre for Precision Farming is working on tree-climbing robots for harvesting dates. The English are pursuing a different concept. Several pairs of wheels connected by joints are tightened around the trunk of the palm tree so that a firm hold is created by friction between the airfilled tires and the tree trunk. Using wheel and disk drive, the robot can climb 30 meters up the trunk. This prototype is also still a long way away from becoming an economically viable product.

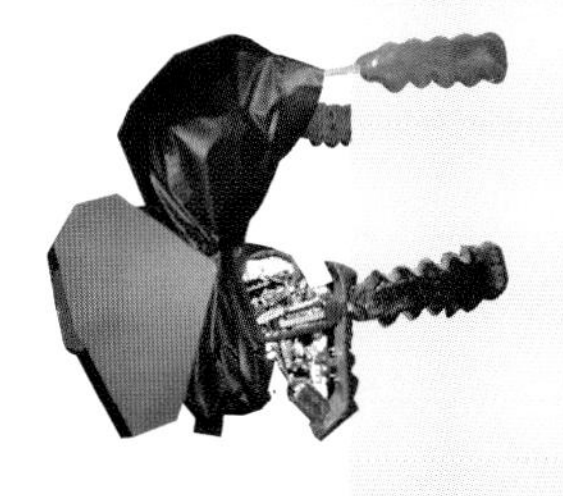

Tree-climbing robot
Peter Brenner,
Germany

Robot YAS1000DX, Yanmar, Japan

Robots in the Dairy

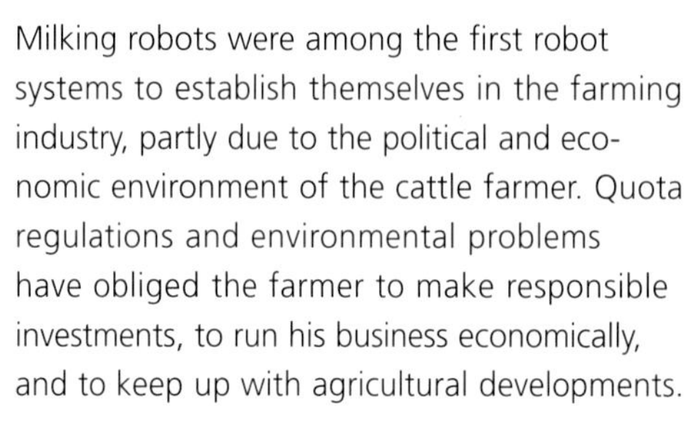

Milking robots were among the first robot systems to establish themselves in the farming industry, partly due to the political and economic environment of the cattle farmer. Quota regulations and environmental problems have obliged the farmer to make responsible investments, to run his business economically, and to keep up with agricultural developments.

The Dutch companies Lely and meko holland bv are the best-known manufacturers of automatic milking systems. The basic functions of Lely's Astronaut and meko's AMS are very similar. The development and success of these two robot systems can be attributed to the fact that cows are creatures of habit and quickly adopt certain behavior patterns. As a result, the majority of the herd has no problem being milked by a robot.

The urge to be milked, as well as some fodder concentrate in the milking box, make the cow go to the robot. The cow wears a transponder collar that is read upon entering the box. A record of each animal is in a database where important data such as milk yield, the number of visits to the robot, and so on, are recorded. If a cow only comes for fodder, even though it was milked half an hour before, the computer system recognizes this fact and the cow goes away hungry.

If the right cow is in the box at the right time, the milking sequence begins. First, the teats are cleaned. A laser scanner then detects the teats, and the milking beaker integrated into the robotic arm is affixed. During the actual milking procedure, the flow, quality, and other parameters of the milk are checked electronically. Once the milking procedure is over, the automatic milking system is thoroughly cleaned, and the animal may leave the box.

- **Automatic milking system AMS, meko holland bv, Netherlands**
- **Milking beaker, Lely, Netherlands**
- **Milking box**
- **Milking robot Lely Astronaut**

Harvesting the American Way

Farming utility vehicles travel a good 1.6 billion kilometers annually in the United States, at speeds of less than 15 kilometers per hour. Driving these farming machines is uncomfortable due to noise and high temperatures, and may be dangerous. Unmanned robotic applications used in this situation promise to double the harvest yield. Autonomously navigated, freely moving robotic systems can work 24 hours a day, seven days a week, and adverse weather conditions have little impact on them.

The National Robotics Engineering Consortium (NREC) came into existence in 1995 as a result of a joint effort of Carnegie Mellon University in Pittsburgh, Pennsylvania; the town of Pittsburgh; and the State of Pennsylvania.

Unmanned farming machine, Institute of Agricultural Machinery (BRAIN - IAM) Saitama, Japan

The NREC, along with the farming machine company New Holland, studied the implementation of an autonomous hay bine harvester within the scope of its first project, Demeter.

One of the standard video cameras mounted on the farming machine records the field of vision in the direction of travel and transfers this information to the onboard computer. The video picture is then divided into subsegments, which can then be classified into cropped or uncropped lines.

The edge of the previous reaping operation is seen as the dividing line, and the current working sequence uses this as a guide. The computer then steers the farming machine in such a way that the cutting line is placed correctly; the cutter bar, of a prescribed width, then dips accurately into the section to be reaped. The prototype, developed within the scope of the Demeter project, has already been thoroughly tested. It performed its job autonomously for 18 hours in Southern California.

Japan is also developing unmanned farming machines to work in the fields. At the Institute of Agricultural Machinery (BRAIN-IAM) in Saitama, various navigation systems have been tested for autonomous farming machines. The researchers have come to the conclusion that, with the present level of technology, autonomous machines can offer approximately the same degree of quality and performance as conventional machines in field cultivation. In the eyes of experts, the reliability of the freely moving machines is not yet high enough, so these machines will not be implemented extensively until the mid-term future. The first step could be one employee monitoring the work of several autonomous freely-moving farming machines simultaneously, a model that German agricultural engineers also are considering.

Autonomous hay bine harvester Demeter, NREC, USA

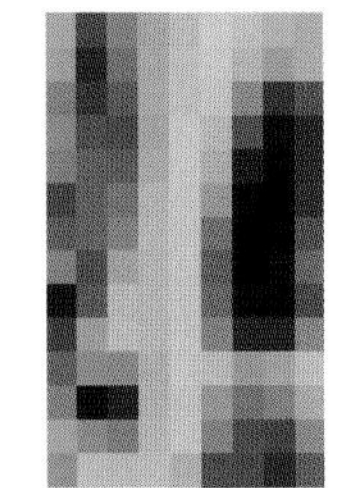

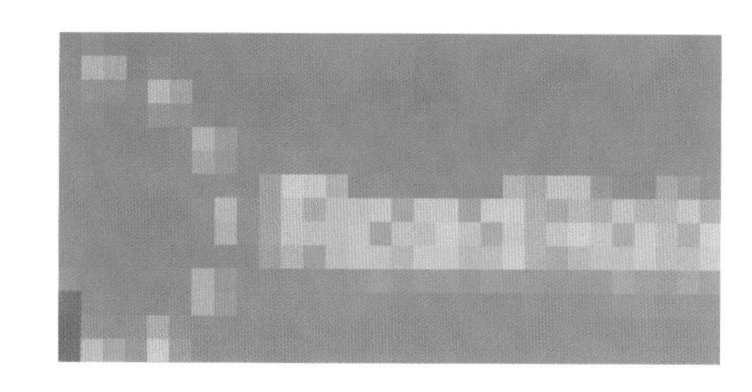

The Construction Industry

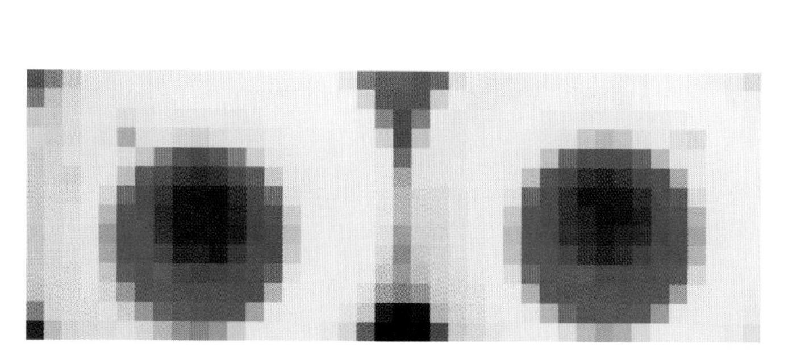

Positioning of exterior facings

Japan's Skyscrapers Originate in Factories

Nowadays, construction machines for performing complex tasks already have been automated to a high degree. A mobile construction robot can be made only by integrating sophisticated sensors and intelligent control technology. In Europe, this technology is not very common.

In Japan, however, robots have been developed for the construction industry since the 1980s. Sponsored by the Ministry of International Trade and Industry (MITI), robots have been constructed to erect houses, to perform dangerous tasks by remote control, to smooth concrete, and to clean building facades. Numerous other examples confirm Japan's pioneering role in construction robotics.

Because basic building conditions in Germany differ from those in Japan, it has not been possible to develop a highly automated construction industry in Germany, but early research shows that German construction machine manufacturers are meeting competition from the Far East head on.

Japan has developed partially automated systems capable of raising skyscrapers floor by floor. Examples of this are the T-UP system from the Taisei Corporation, the Smart system from the Shimizu Corporation, and the Akatsuki 21 from the Fujita Corporation.

Construction Without Dirt and Noise

Akatsuki 21 is made up of three segments: The ground factory, the transfer line, and the sky factory. The ground factory is a stationary system positioned in the basement of the building to be erected, and is responsible for unloading building materials as they arrive, commissioning them, and then transporting them to the transfer line. To simplify the final assembly of the construction components used, the ground factory also pre-assembles building materials.

In this way, assembly time in the sky factory is reduced to a minimum. Autonomous Guided Vehicles (AGVs), loaded and unloaded by a portal crane, take over the transportation of double-T beams, facing elements or staircase segments. Supply units for the welding equipment and several concrete pumps are also found in the ground factory.

The building materials required are forwarded via the transfer line to the sky factory. The chief component of the transfer line is a goods lift which picks up the loaded autonomous transportation systems from the ground factory and transports them to the desired floor. The lift cabin has a floor surface area of 3 meters by 11 meters, and is capable of transporting up to 10 tons.

The sky factory performs the actual construction tasks. It is assembled above the ground factory before building commences and contains all the machinery and equipment necessary for completion of the construction.

The Roof is Built First

The sky factory is basically a type of hydraulic ramp with a factory integrated into it. Once one floor of the building has been completed, the sky factory moves exactly one floor upwards.

The hydraulic cylinders, responsible for lifting the entire sky factory segment upwards, are housed in the supporting columns of the building itself. The difficulty in moving the top floor lies in activating all the hydraulic cylinders simulta-

Skyscraper Akatsuki 21, Fujita, Japan

neously. The lifting procedure takes place every eight days. Once the construction work is over, the sky factory is taken down and dismantled. All tasks are coordinated and controlled by a control center in the sky factory segment.

Robots can Build in All Weather

There are many advantages to building skyscrapers this way. By using the sky factory concept, the entire construction is protected from the influence of bad weather right from the start because the top floor is built first. The construction workers are able to carry out their tasks under the best possible conditions because they are always in a closed building. Neighbors also welcome this form of construction, as noise and dirt rarely occur on building sites of this nature.

The initiators of construction projects also have significant advantages. In comparison with conventionally built skyscrapers, construction time is reduced by 30%. Akatsuki 21 automates a large part of the tasks necessary for building skyscrapers. Included here is work on the steel structure and the positioning of exterior facings. Human error cannot occur with this form of construction.

Akatsuki 21 can be considered a "mobile factory for automated construction".

Inside Akatsuki 21, Fujita, Japan

Sky factory Akatsuki 21, Fujita, Japan

Unmanned Telemanipulators Under the Volcano

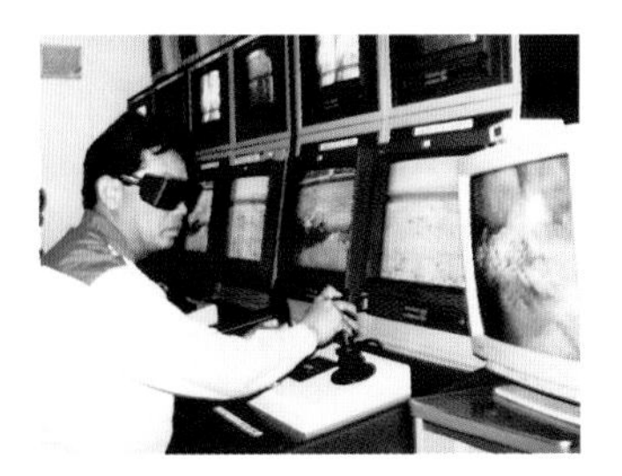

Control station for teleoperation

Telemanipulation is an important issue in Japan's construction industry. Since November 1991, Mt. Fugen, a volcano near Nagasaki, has been active again and has claimed the lives of over 40 people. Consequently, the Japanese Ministry for Construction invited bids for a project for the removal of erupted lava masses and mountains of ash by telemanipulation. The Fujita Corporation was awarded the contract. As a result, four completely unmanned construction sites with dump trucks, diggers, and bulldozers were created.

Manipulators and video equipped robots perform the more complex tasks. All equipment is remote-controlled from a control station 1.8 kilometers away. The operator, thanks to 3-D projection, has the impression that he is in the immediate vicinity of the remote-controlled units. A satellite-aided global positioning system (GPS) enables the operator to locate his machines exactly on the construction site.

An automatic measuring system continuously calculates the amount of earth to be excavated. Amounts excavated to date on the remote-controlled construction site are comparable to those attained on standard construction sites, as substantiated by statistics from the Japanese Ministry for Construction. The average quantity of earth excavated by an operator in a vehicle is 2,260 cubic meters per day; amounts excavated on the experimental grounds lie between 1,500 and a maximum of 2,750 cubic meters per day.

For Ramparts

The latest research is concerned with building a rampart made of concrete blocks by teleoperation. An unmanned manipulator picks up concrete blocks from the loading area of a dump truck, which is also unmanned, and assembles them to make a concrete wall. The first tests were successfully carried out in the spring of 1998.

Unmanned construction site, Fujita Corporation, Mt. Fugen, Japan

Telemanipulated Mini Digging Robot,
Tokyu Construction Co. Ltd., Japan

Automated Digging

The Mechatronics Research Department of the Tokyu Construction Company is also carrying out research in the field of telemanipulated construction machines. In connection with this research, the Mini Digging Robot was developed. This is a remote-controlled digger with an earth borer integrated into it. The construction machine weighs 2.13 tons and requires a three-phase alternating current with a power demand of 20 kilowatts. Hydraulic engines are used for movement.

The teleoperated construction machine is capable of excavating 6.9 cubic meters of earth per hour. The ground is loosened using an earth borer integrated into the digger so that the excavation procedure with the dredging shovel is easier. The Mini Digging Robot can cope with an earth compression of 50 kilograms per square meter. The title "Mini Digging Robot," chosen by the manufacturer, is not strictly correct, as this construction machine is only a remote-controlled digger and not a true robot.

Climbing Robot that Cleans

The Tokyu Construction Company has also developed "real" service robots such as the Wall Surface Operation Robot. The wall climber makes contact with the facade using two vacuum grippers and is guided by two cables.

Adapted onto a carrier system, the Cartesian kinematics, with their tool-holding fixture, have a positioning accuracy of +/- 0.5 millimeters. The Wall Surface Operation Robot weighs 80 kilograms and can position a payload of up to 10 kilograms on a wall with an accuracy of ±5 millimeters.

The climbing robot's job is to check, clean, and paint the facades of buildings. The paintwork may be of more than one color; the robot is capable of painting a pattern in ten different colors. The accuracy of the Wall Surface Operation Robot is especially important when working on patterns.

The surface performance of this service robot attains approximately 250 square meters per day in the case of inspection and measuring tasks. When using a steel brush for cleaning, performance drops to approximately 150 square meters per day. If only one color of paint is applied, a wall surface area of 670 square meters can be painted in a day. If patterns are to be painted, the robot can only achieve a maximum of 55 square meters per day.

The system can be utilized only on smooth surfaces; balconies or projections make use of a cable system for vertical movement impossible.

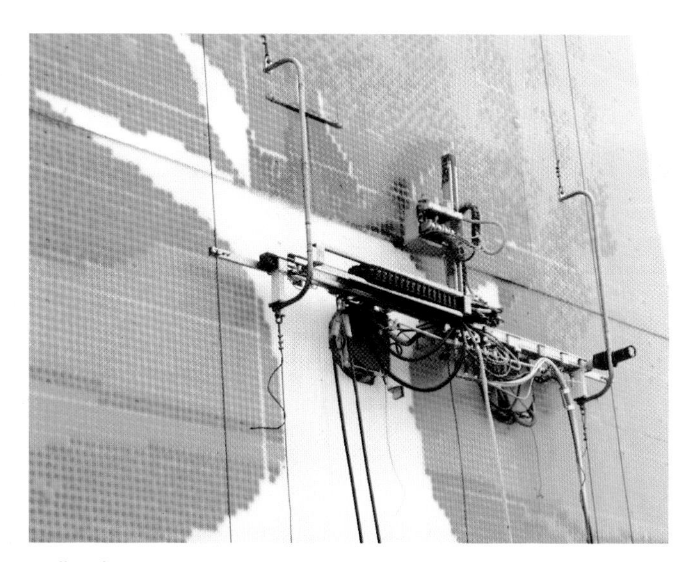

Wall-Surface Operation Robot,
Tokyu Construction Co. Ltd., Japan

Floor Troweling Robot, Hazama Corporation, Japan

Coating Unit for Concrete Floors

Another example of Japan's ingenuity in the field of construction robots is the Floor Troweling Robot from the Hazama Corporation. This autonomous vehicle is used to trowel off freshly poured concrete floors.

The first procedure, the rough troweling off and leveling of the concrete surface, still has to be carried out manually. The labor-intensive and physically tiring job of fine troweling, however, is taken over by the robot.

The modularly constructed Floor Troweling Robot can be dismantled into such small pieces that it can fit into the trunk of a car and can therefore be easily transported. Assembly takes just ten minutes. Once assembled, the service robot's dimensions are 1,990 by 800 by 910 millimeters and it weighs 100 kilograms.

Floor Troweling Robot fits into the trunk, Hazama Corporation, Japan

Teach-in Before Troweling Off

The mobile robot trowels off the concrete surface along a predefined path that is programmed into the system in teach-in mode using a remote control. Once the path has been memorized, it can be traveled automatically.

teach-in mode
On-line programming is accomplished through teach programming methods. Teach programming is a means of entering a desired robot control program into the robot controller. The robot is manually led through a desired sequence of motions by an operator who is observing the robot and robot motions as well as other equipment within the workcell. The teach process involves the teaching, editing, and replay of the desired path. The movement information and other necessary data are recorded by the robot controller as the robot is guided through the desired path during the teach process. (Handbook of Industrial Robotics, 2nd Edition, 1999 John Wiley & Sons)

The mobile robot is able to recognize unforeseen obstacles with the aid of distance sensors positioned at the front and rear in the direction of travel. Should the system fail for any reason, the robot is still able to detect obstacles with its tactile fenders, which are also located at the front and rear of the vehicle.

When the concrete starts to harden, the robot is placed onto the surface and first smoothes the floor with a sort of wooden trowel driven by an eccentric motor. The robot is then removed from the surface to be worked on so that the building materials may recover a little. In the second stage of the procedure, the Floor Troweling Robot is placed onto the surface and, using a metal trowel oscillating at a frequency of 23.5 kHz, gives the surface its final smoothing.

The mobile robot is equipped with a built-in energy generator, guaranteeing an autonomous energy supply.

Europe Is Building Roads with "ESPRIT"

The RoadRobot is the first road finisher in the world to be self-navigating, self-steering, and able to perform all the necessary tasks to make a road surface meet all desired requirements, from distributing materials, leveling, and profiling to even compressing, and all computer-controlled.

The European Union sponsored the project within the scope of the ESPRIT program. Six companies from five European countries (Germany, England, Holland, Portugal, and Spain) were involved. Together with Joseph Vögele AG, Mannheim, Germany, the European Center for Mechatronics in Aachen was in charge of the project.

The aim of this European Union project was to engineer a work device for road engineers to allow them to build road surfaces more consistently than before, and in a more environmentally friendly way.

What Happens When a Road Is Built?

The processes of a road-finisher can be divided into four subsystems: the logistics of coated materials, traveling mechanism, geometry of the road surface, and the plank. Each subsystem has its own process module. The logistics of coated materials means in this case the receipt of coated materials from the truck, their passage through the finisher towards the rear, and their correct distribution in front of the plank.

The traveling mechanism includes the speed control, start/stop function, and steering of the finisher. The surface geometry module takes into account the thickness, profile, and width of the surface and also any lateral inclination. Curbstone edges or leveling to the same height of existing road surfaces are also considered. Everything to do with a plank, such as tamper, vibration for surface compression, pressing ledges, plank heating, and plank relief are integrated into the plank subsystem.

The RoadRobot is controlled either by radio from the engineering office or via the on-board computer touchscreen. All information concerning line direction determination, route level, and the bed to be surfaced (a layer of crushed stone, for example) and also the handover of coated materials, is determined by a well-thought-out sensor system, and data is transferred by a process module to the onboard computer. The control and regulating processes are shown on the central onboard computer's touchscreen and can also be adjusted here.

Diesel-Electric Drive Is Environmentally Friendly

The RoadRobot's diesel-electric drive was developed by Vögele AG together with Kaiserslautern University. With this drive, the RoadRobot is up to 12 decibels quieter, produces 50% less exhaust fumes, and requires 50% less fuel than conventionally-driven road finishers with an equal performance level.

The electrical drive requires no hydraulic oil. In this respect, the utilization of this road finisher in nature and water reserves is not hazardous, and no costly safety precautions need be taken either.

Road Robot, Joseph Vögele AG, Germany

No-Dig Line Construction

Ever since Roman times, man has been tearing up the earth to meet his needs. Today, we lay cables and pipes for water, drainage, and gas. Since 1986, however, a European alternative to this has been in existence: line construction using a horizontal boring technique without the need to dig trenches. The FlowTex company in Karlsruhe was, and still is, the pioneer of this technology.

The horizontal boring process is a fully controllable, gentle, and environmentally friendly wet-bore technique. The advance work functions on a combined principle. Instead of using conventional mechanical boring techniques, fine, sharp-cutting water jets are emitted from nozzles at the tip of the boring head, which cause the hydro-mechanical "holing-through" of loose rock. The loosened material is removed by the back-flow along the length of the boring rod, and also causes the loose rock to be gently redistributed and compressed with less pore room.

The entire process control of the boring takes place by way of a transmitter built into the lance of the bore; the electromagnetic field created by the transmitter can be located at any time using a field strength measuring device on the earth's surface. The tip of the boring head is flexible and its orientation is used to control the bore lance. In the case of small boring systems, the minimum turning radius is 8 meters, the boring section can be as long as 200 meters, and the maximum depth 8 to 10 meters.

The newest generation of machines have different repetitive work sequences automatically incorporated into them to make the machine operator's work easier, for example, cleaning and greasing the boring rod, or handling the boring rod, including magazining. Due to the partially automated work sequence control, the machine's boring performance could be improved, and an even strain on the boring rods ensured.

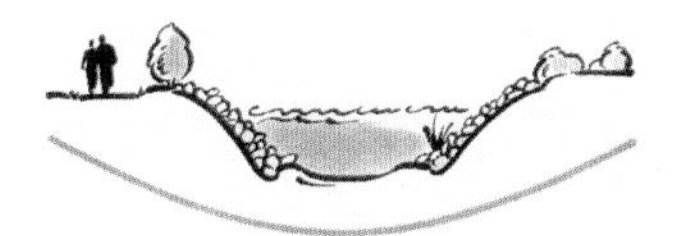

Drilling-robot, FlowTex, Germany

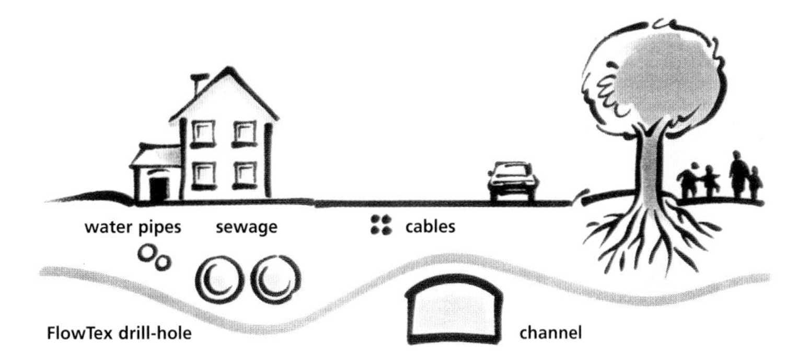

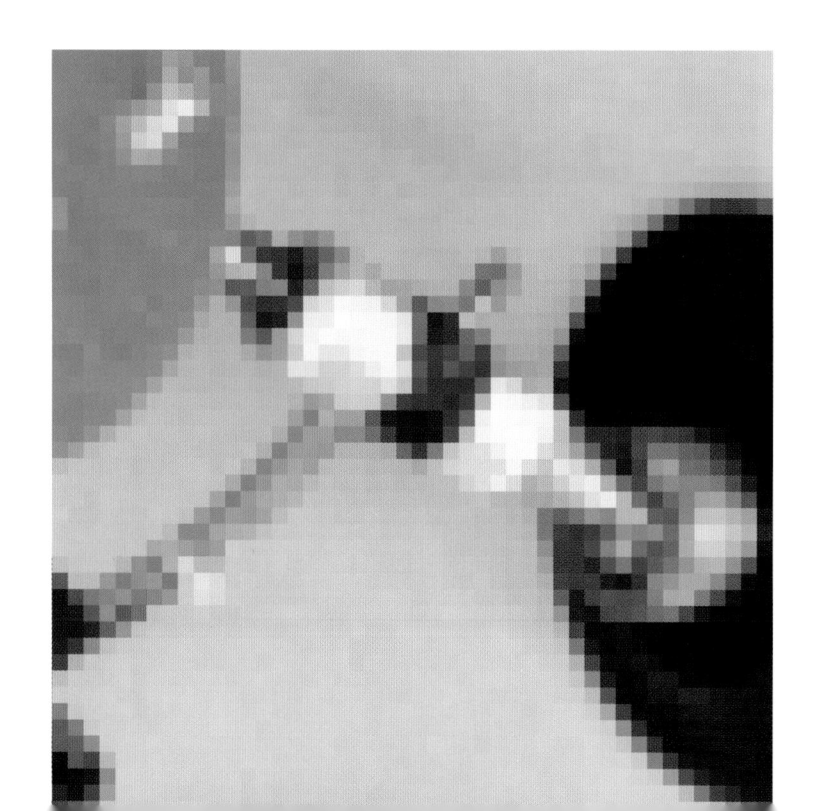

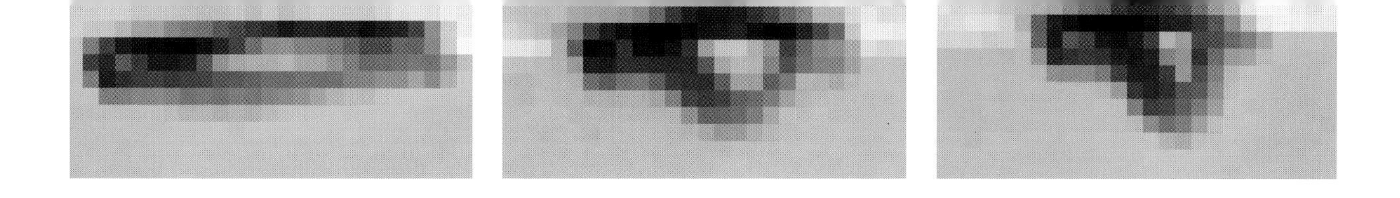

Renovating

Inspecting and Renovating in Hazardous Environments

Humans have their natural limits. Our working environments, however, often force us to go dangerously close to these limits. Tasks such as firefighting, working at great heights, performing cleaning jobs in the toxic air of a chemical tank, or carrying out renovating tasks in the radioactively charged core zone of a nuclear power station are all associated with great danger and enormous health risks for human workers.

Even when all safety regulations are followed, a certain risk remains. The economic feasibility of jobs which can only be carried out by individuals wearing protective clothing and, for safety reasons, only for a very limited period of time, is also questionable.

New types of service robots that are able to move around in hazardous environments independently, safely, and with ease are required: robots that do not get tired or overtaxed on multiple-shift jobs and, if a catastrophe does occur, do not leave behind a grieving family, but rather an insurance claim.

Robots in Dangerous Situations

As an example, we need autonomous climbing robots for renovating large buildings. Ever since the sixties, concrete has become indispensable for constructing bridges and dams, but concrete does not last forever. In recent years, an increase in polluting gases and harmful substances has added to the natural stresses of humidity, temperature fluctuations, and solar radiation.

These stresses cause chemical reactions to occur on the surface of concrete constructions. Cracks form, leading to corrosion of the reinforcing steel near the surface. This puts the stability of such constructions in grave danger. Another problem, which we have learned about from highway studies, is that a large part of the damage results from faulty construction.

Autonomous climbing robots are particularly suitable for undertaking the work in renovating such damage. Using human workers here would be far too dangerous, too costly, and even technically impossible in some cases.

Nuclear Power Stations Have High Renovating Costs

What is not generally known is that nuclear power stations have enormous renovation costs. There are 19 active nuclear power stations in Germany alone; worldwide, there will soon be 500. A check needs to be carried out at least once a year and can take several weeks. Each day that a nuclear power station remains shut down costs the energy supplier over half a million dollars. Reorganizing this hazardous work by implementing robots is soon worth the money.

There are also 16 non-operational German nuclear power stations. Sooner or later they will have to be dismantled. If we did not know before, we have certainly learned since Chernobyl how dangerous atomic ruins can be. Humans are too valuable to perform transportation, maintenance, or inspection tasks. Service robots are largely resistant to radiation, are reliable, and are almost tireless.

autonomous
Greek: "living under one's own laws", self-reliant, independent

bionics

The word "bionics" comes from a combination of the two terms "biology" and "electronics". A characteristic feature of bionics is their interdisciplinary relationship. This young and very promising research area combines biology with engineering sciences, and also with architecture and mathematics. The aim of bionics is to transfer nature's solutions to problems to the world of technology in order to capitalize on "nature's discoveries" which have been developed and optimized over millions of years. The following definition is recognized by leading specialists in the field of bionics: Bionics as a branch of science is systematically concerned with the technical transfer and application of constructions, processes and the developmental principles of biological systems. (Neumann 1993)

Adhesion

- The holding on of two materials or bodies to one another
- The holding on of molecules to each other at the outer boundaries of two different materials (glue)

How a Climbing Robot Functions

A climbing robot is a mobile platform that transports a tool or sensor to a place that is hazardous or inaccessible to humans. The mobile platform has to do two things: It must adhere securely to the underside of a surface, and must also be able to move at the same time.

Nature has shown us how to hold on to the underside of surfaces: either by form closure or an adhesive force. Flies have small hairs which hook on to a microscopically rough surface, and in this way grip using form closure. When in danger, small beetles secrete a liquid that creates an adhesive force between the underside of the surface and the insect's legs, resulting in a connection with the ground. An attacker is no longer able to turn the beetle onto its back. Squid are equipped with suckers. The examples go on and on.

Frictional Connection and Form Closure Give a Strong Hold

If a construction has a profile, tracks, or something similar, form closure is the gripping method recommended to give climbing robots a secure hold. If such aids are not available, contact to the underside must be established using a frictional connection.

In the case of the steel surfaces of gas storage tanks, petrol tanks and ships' hulls, a grip can be achieved using electromagnets. Vacuum grippers use negative pressure to hold on to concrete walls.

Bionics - Examples Taken from Nature

The motion apparatus of a robot may be designed either as a caterpillar system, a multi-legged frame, or a so-called sliding frame.

A sliding frame is made up of a minimum of two platforms that can be displaced against one another. Each platform is equipped with its own gripping system. While the gripping system of one platform ensures a firm hold on the underside, the other platform is moved forward. The gripping system of the platform that was shifted then takes over the grasping function, and the first platform can also be advanced.

Caterpillars, Beetles and Spiders

We know about caterpillar systems from tracked vehicles. In addition to its caterpillar system, a climbing robot also needs a gripping system.

Multi-legged creatures are modeled by nature. Climbing robots may have four, six, or even eight legs. There are no limits to creativity with robots. Here as well, each leg is equipped with an element for adhering to the underside of surfaces.

Depending on the tasks required, the type of working system mounted onto the mobile platform may vary: Climbing robots guide video cameras, carry actinometers and other sensors, or are equipped with tools, such as drills or right angle grinders. The forces created by using the tool determine the requirements of the adhering system. It must ensure a firm hold and an adequate antagonistic force in even the toughest situations.

Creation of an adhesive force

F

Adhesion by form closure

Gripping system; F = force

Caterpillar system

m

Multi-legged system

m

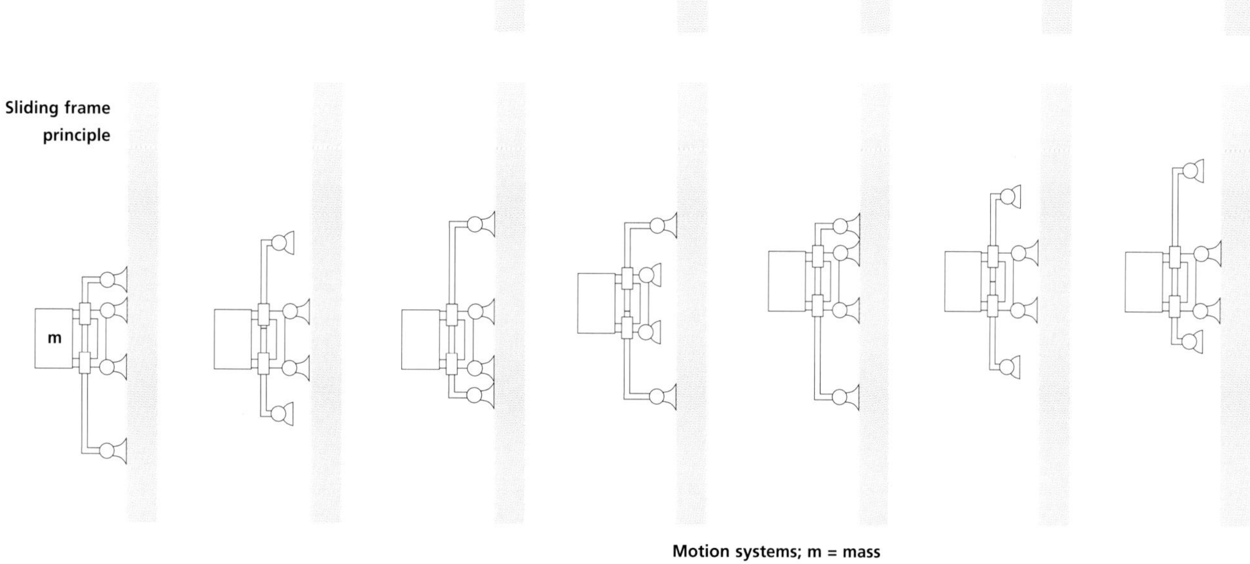

Motion systems; m = mass

TRIBOT, Portech, GB, equipped with a water gun

Climbing Robots in Use

Portech Ltd. is among the leading manufacturers of autonomous climbing robots in the world. The English company is based in Portsmouth and works in close cooperation with the University of Portsmouth. Portech counts the chemical industry and managers of nuclear power stations among its customers.

The prototype ROBUG IIa is a research object for Portech's design engineers. The spiderlike robot has four legs, each equipped with suckers connected in pairs via a turning knuckle at the center of the system. Step by step, ROBUG IIa advances; due to its lower center of gravity, it has a very stable gait. It is even capable of leaving the ground and climbing up a wall.

The Cleaning Service can Climb Walls

Portech's TRIBOT is based on the sliding frame principle and is equipped with a swiveling water jet nozzle. It undertakes cleaning duties on large surfaces outside the nuclear power plant industry. TRIBOT adheres to the underside of a surface by using specially developed vacuum grippers that give a firm hold even on rusty surfaces.

SADIE was developed for the British company Magnox Electric and is the youngest member of Portech's robot family. This climbing robot carries out inspection work on CO_2 flow channels in nuclear power stations equipped with Magnox-type pressure vessels. SADIE is also based on the sliding frame principle and uses vacuum grippers to hold on to the underside. Six grippers are used, but only two are really necessary for this device to adhere to a vertical wall.

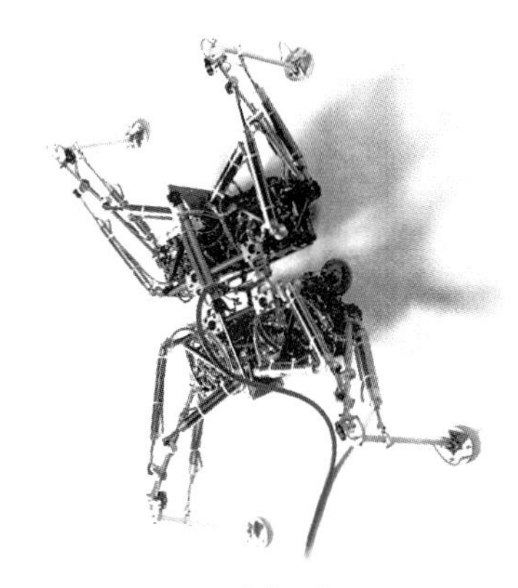

Robug IIa

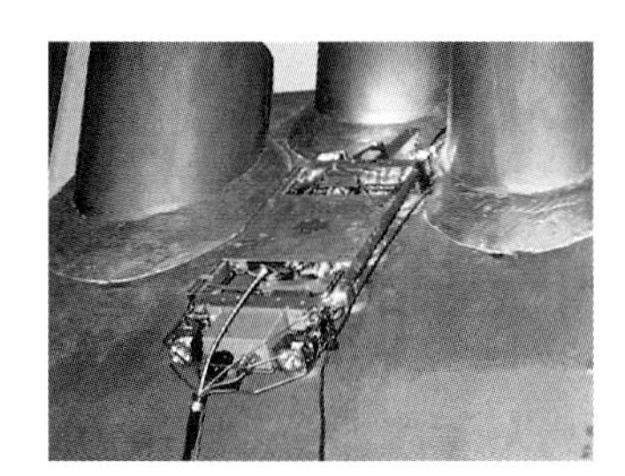

NERO inside a nuclear power plant

Robug III together with Prof. Arthur A. Collie, Technical Director, Portech Ltd., GB

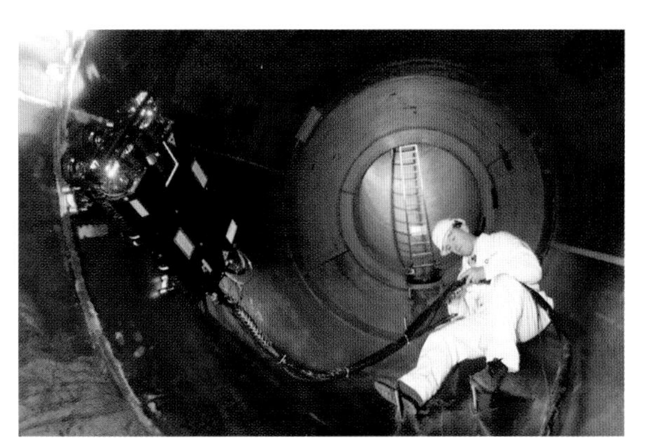

SADIE, Portech Ltd., Great Britain

RoSy I carrying out an inspection on a highway bridge, Yberle Robotersysteme GmbH, Germany

HYDRA II

RoSy Takes a Temporary Job on a Building Site

The climbing machines RoSy I and RoSy II come from the Robotersysteme Yberle GmbH in Neumarkt/Oberpfalz in Germany; RoSy stands for "Remotely Operated System Yberle". Equipped with various tools, these teleoperated climbing robots can undertake tasks in inaccessible areas of the building or carry out construction maintenance. They can perform various functions, such as removing materials, painting surfaces, drilling holes, and putting in wallplugs, as well as carrying out general checking, inspection, and cleaning tasks.

RoSy uses vacuum grippers as its adhesive system. The motion apparatus, a form of kinematics derived from the sliding frame principle and comprised of lifting elements and moving units, enables RoSy to move freely in all directions.

Testing of RoSy II

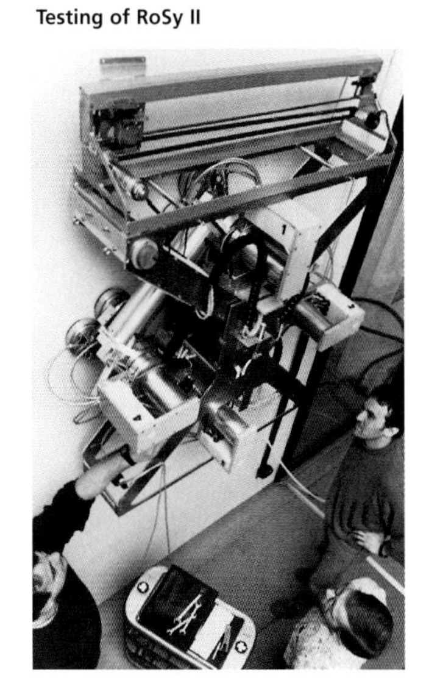

Yberle Robotersysteme GmbH, Germany

HYDRA

The HYDRAs, developed by the UWTH at the Institut für Werkstofftechnik at Hannover University, Germany, can cope with any working position and any tool. The HYDRA carrier is a task-adapted kinematic system that uses vacuum grippers and weighs between 10 and 38 kilograms; depending on the model, it can transport payloads of between 5 and 50 kilograms.

HYDRA IV

HYDRA V, UWTH, Germany

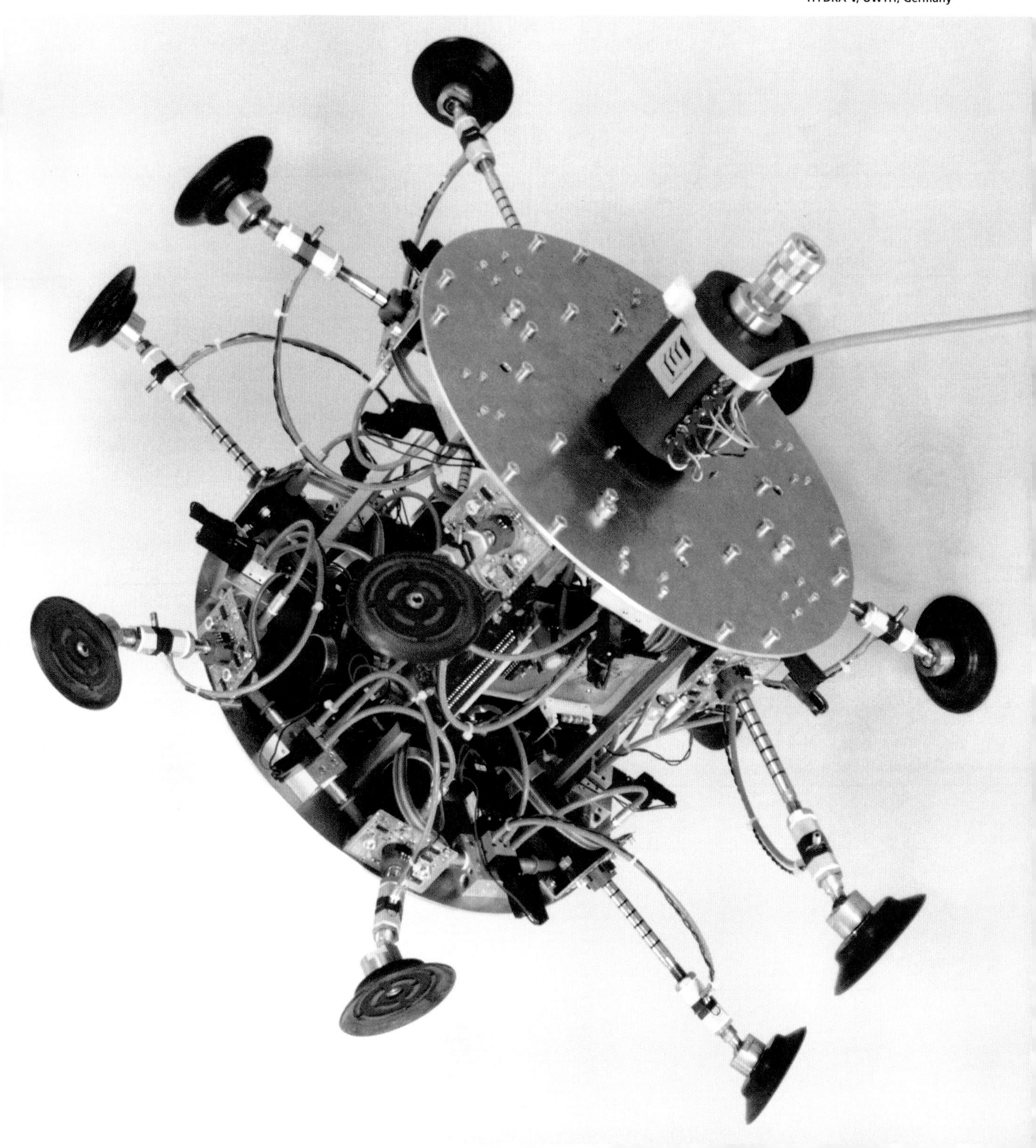

Development Trends

Nearly all known climbing robots to date use active suction elements, electromagnets, or a combination of the two to grip onto the underside of surfaces. Energy, compressed air, and electricity are all supplied by a fixed ground station via an umbilical.

The finite length and weight of the supply cables limits the range of these robot systems. The cables can also get caught on obstacles in deeply fissured working areas, making it more difficult to plan the robot's path of travel.

Adherence Without Requiring an Energy Supply

At Fraunhofer IPA in Stuttgart, Germany, a gripping system for climbing robots is being developed that uses passive adhering elements and thus needs no compressed air lines. On particularly smooth surfaces, such as glass or structural glazing facades, a vacuum can be created passively inside the gripper by simply expelling the residual air in the elastic suction cup. The amount of air penetrating between the gripper's sealing lips and the underside is so small that an adequate negative pressure can be maintained for a sufficient length of time without having to constantly pump air out of the evacuated suction foot.

Components and mode of operation of the valve mechanism

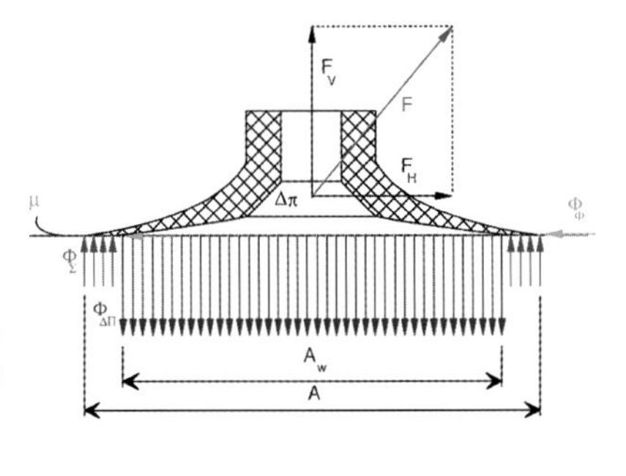

Distribution of force inside the gripping element

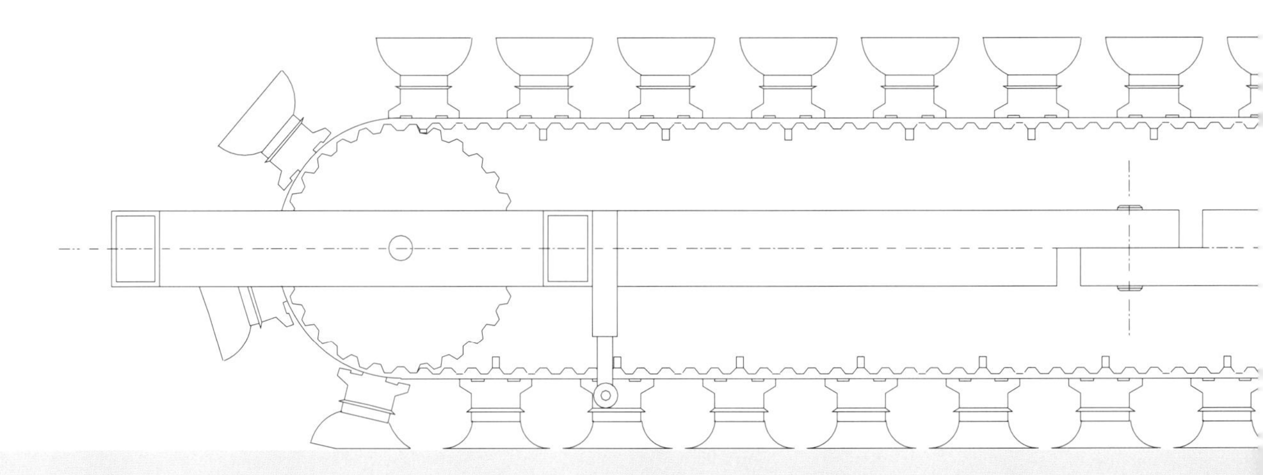

Millipede Can Climb Up Glass Surfaces

"Millipede" is the name of the first prototype equipped with a passive gripping system. It establishes contact with the underside via 60 suction cups evenly arranged in pairs along the surface of a toothed belt.

Each of the caterpillar system's 60 gripping elements is equipped with a valve mechanism, and this patented invention performs three tasks. When the suction cup is pressed onto a surface, the valve opens and the air inside can escape. Once the air has been released from the area covered by the suction cup, the valve closes and prevents the equalization of pressure with the atmosphere. The adhesive strength of the gripping element is then at a maximum.

On detachment, the valve allows air into the gripping element and the sucker's adhesive power is reduced to zero. The valve mechanism is controlled purely mechanically by the synchronized shafts and tension rollers of the toothed belt's drive.

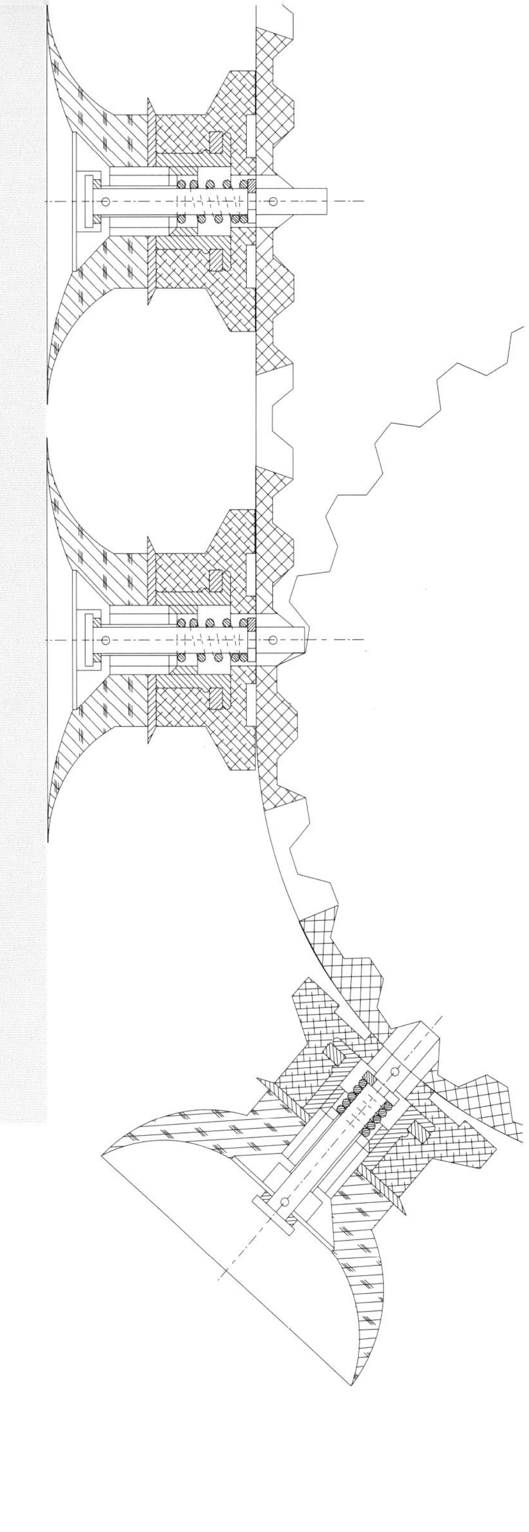

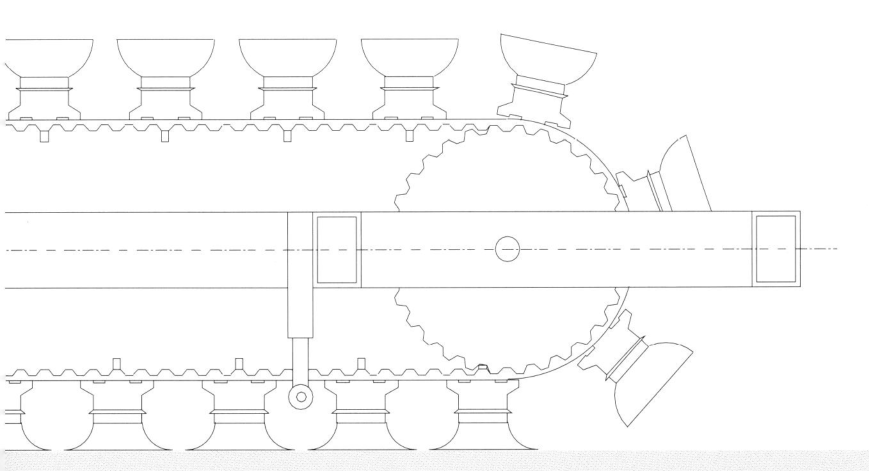

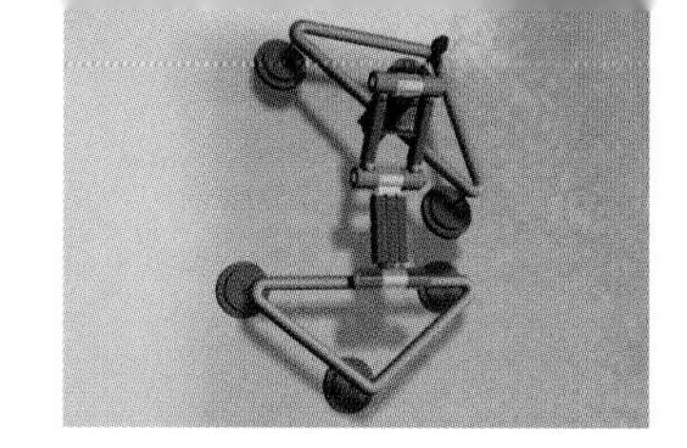

SOLIST - Caterpillar and Performer at the Same Time

Aided by suction grippers, SOLIST, a lightweight carrier for inspection systems able to go just about anywhere, was conceived within the scope of a study at the UWTH at Hannover University, together with the Swiss company Colenco Power Engineering AG in Baden. SOLIST has also been designed to perform maintenance work in areas inaccessible to humans.

Tools required for carrying out such maintenance tasks and which can be adapted to the system are, for example, abrasive cutting tools, small manipulators, measuring sensors, welding wrenches, welding torches, cleaning devices, and painting units.

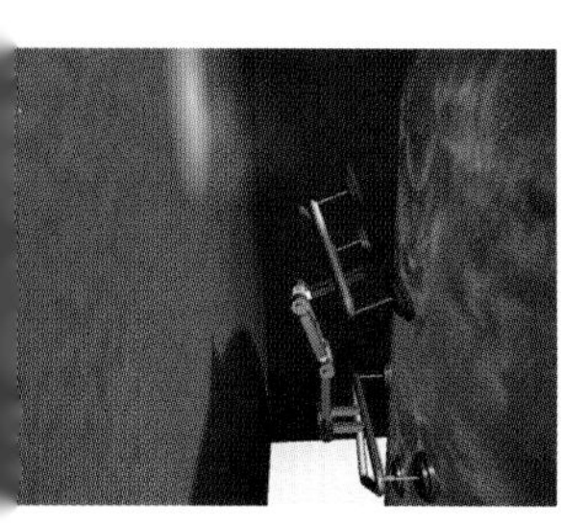

Enormous demands are made on the climbing robot's kinematics. It has to be maneuverable and be in a position to climb over or pass beneath all sorts of obstacles. The kinematics of the robot must also enable it to walk on one- or two-dimensional curved surfaces, such as a sphere or the outer wall of a cylindrical object. It must also be able to change its plane of travel at any time, as it does when it leaves a wall and climbs onto the ceiling of a room.

Because kinematic concepts used before the present time have been unable to fulfill such demands satisfactorily, a completely new movement mechanism has been designed. This new system consists of two triangular platforms, each equipped with three suction feet. Both these platforms are connected via an articulated arm to six electrical drives that can be accurately controlled.

The arm has rotatable drive units in the middle and at each end, and can fully extend or fold into an M. The length of the individual arm limbs is also variable. The robot is even capable of changing direction on the spot by turning on a standing platform or rotating around its stabilizing foot. The system's kinematics also enable it to move like a caterpillar.

Examples of use and forms of motion of SOLIST

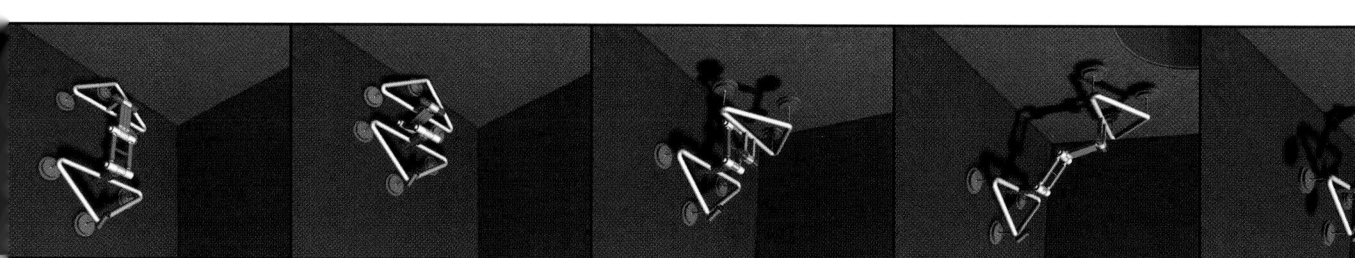

SOLIST

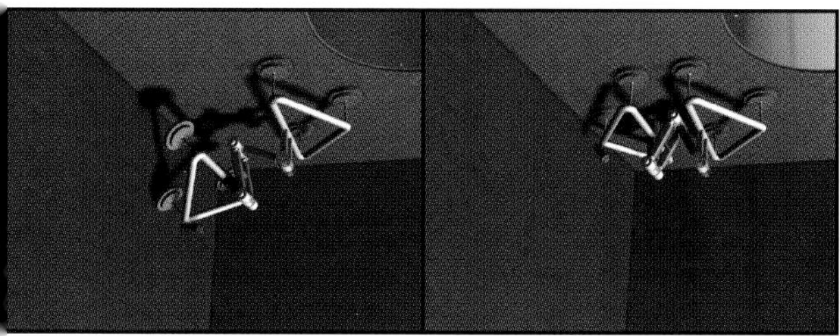

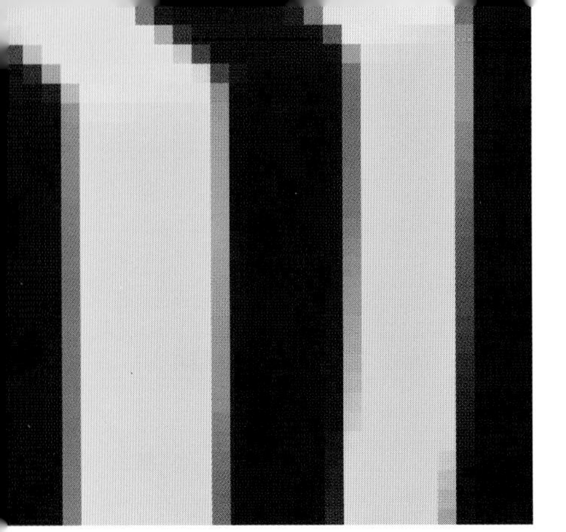

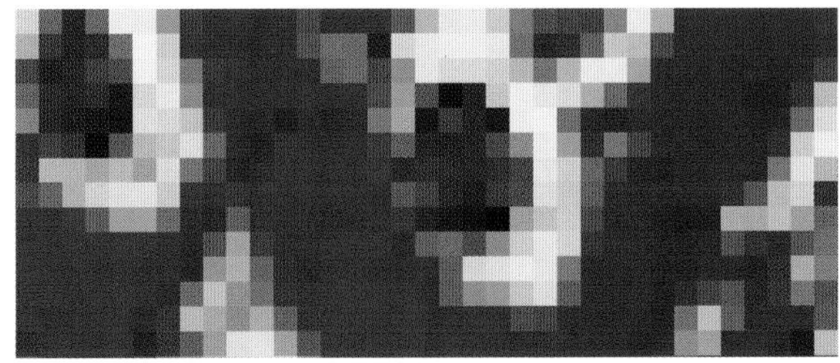

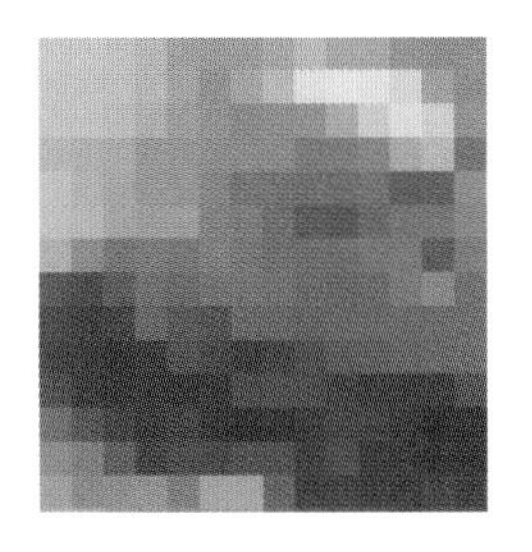  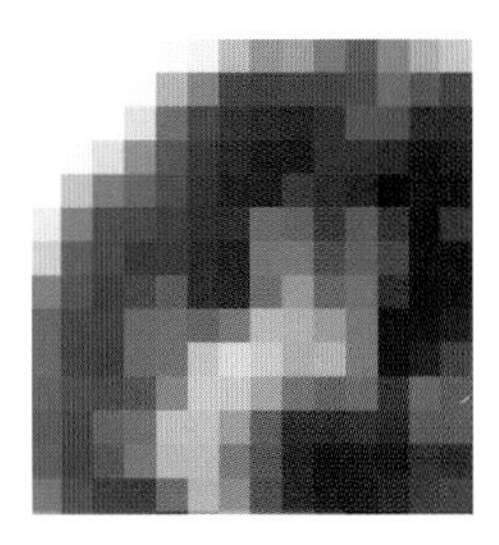 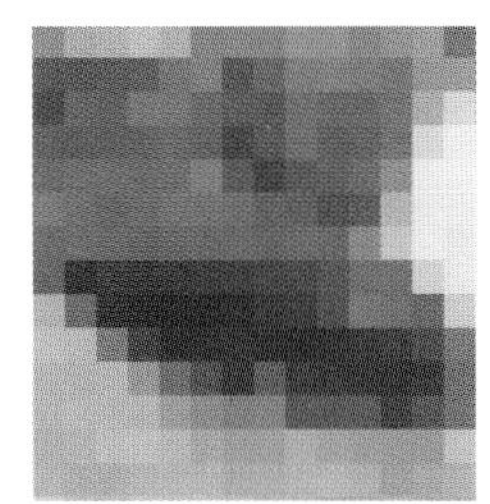

Cleaning

Cleaning Floors

Cleaning Facades

Cleaning Aircraft

Cleaning Recreational Boats

Cleaning Floors

Floor-cleaning robot, Hitachi, Japan

Automated Floor Cleaners

Cleaning is a difficult and poorly paid job. Whole armies of cleaners, busy every single evening sweeping, vacuuming, spraying, wiping, waxing and polishing, are responsible for keeping airports, stations, and office buildings clean.

Wages make up more than two-thirds of the costs incurred from cleaning. For this reason, cleaning companies and manufacturers of floor cleaning machines both see enormous potential for automation in this field.

Autonomous floor-cleaning machines have been developed and sold in Europe, Asia, and America for over 20 years. Such cleaning robots have become a daily sight in Japan Railway's Tokyo and Osaka stations and at the Haneda Airport.

In France, French cleaning companies such as Abilis, Comatec, and GSF have been implementing various cleaning robots since the early 1980s. Commissioned by RATP, the Parisian metro, approximately 25 autonomous cleaning machines from Comatec carry out their tasks. In the Dutch supermarket chain belonging to the Albert Heijn B.V., the autonomous scrub/vacuum machines ST-81, developed by the Siemens AG and the Hefter Cleantech GmbH, can be seen. According to estimates made by the Fraunhofer IPA, between 40 and 100 autonomous floor-cleaning machines are in commercial use in the world to date, not including existing prototypes.

The field of automated floor-cleaning machines is really starting to progress, as shown by the new developments made by Siemens-Hefter, Kärcher, and Minolta. Kärcher and Minolta are interested in the commercial service area and will also consider the domestic area in the near future.

Floor-cleaning robots can be divided into two categories: those intended for the commercial sector and those for the home.

ST-81 inside an Albert Heijn B.V. supermarket

Autonomous scrub/vacuum machine ST-81 , Siemens AG and Hefter Cleantech GmbH, Germany

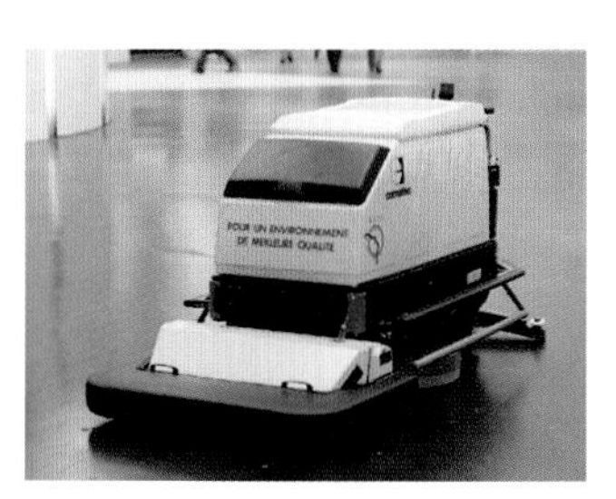

Floor-cleaning robot AUROR, Cybernétix, France

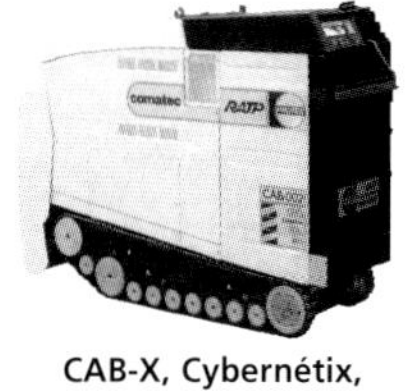

CAB-X, Cybernétix, France

Floor-cleaning robot, Panasonic, Japan

CYBERVAC, Cyberworks Inc., Canada

BR 600, Alfred Kärcher GmbH, Germany

A Household Helper

Compared with established cleaning robots manufactured for the service sector, robots developed for use in the home are built on a completely different scale: homecleaning robots need to be very small, light, and reasonably priced, they must be highly navigable and maneuverable, and they must have a variety of safety features.

The requirements of a floor-cleaning robot are complex. Without running into anyone or anything, it must be able to pass under tables and chairs, reach into tiny corners, and navigate around the expensive Chinese vase. It shouldn't drive the dog crazy and should give the hamster a reasonable chance of survival if paths cross. It must also be simple to program the cleaning system to recognize the arrangement of individual spaces.

Respect for Dogs and Stairs

Engineers from Minolta have challenged themselves to meet these goals. One of the first prototypes weighs 8.2 kilograms, and its dimensions of 321 by 320 by 170 millimeters make it particularly small. Its two driven wheels and slightly rounded shape enable the cleaning robot to get close to walls and into corners. Ultrasound touch and level sensors even allow it to recognize stairs. Its nickel-metal hybrid battery keeps it going for two hours. An Internet connection will enable the cleaning robot to communicate.

In Germany, the Alfred Kärcher GmbH company has won the race to develop the first autonomous floor vacuum cleaner. During the Domotechnica in 1999 in Cologne, Germany, the largest trade fair for household appliances in the world, Kärcher presented the 2000 RoboVac. The robot measures 200 by 200 by 80 millimeters and weighs just 1.4 kilograms. After 20 minutes of hard work, the little helper takes an equally long break, recharges its batteries, and disposes of its dusty load. In this way, it is able to vacuum an area of 15 square meters per hour.

It is clear to all design engineers that there are still some acceptance problems to be dealt with in the home as far as homeowners are concerned, even though the 2000 RoboVac and other similar cleaning robots have enormous potential in the mid- to long-term future when they are accepted.

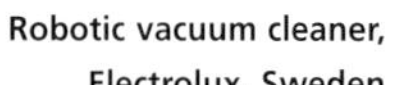

Robotic vacuum cleaner, Electrolux, Sweden

Autonomous floor vacuum cleaner 2000 RoboVac, Alfred Kärcher GmbH, Germany

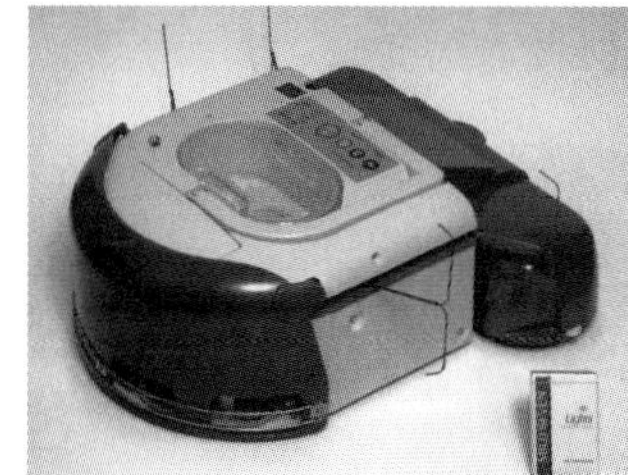

Cleaning Robot, Minolta, Japan

Clean Station Campaign

If the enormous traffic of people on train station platforms and in terminals is taken into consideration, it is no wonder that autonomous floor-cleaning machines have been introduced here first.

The Hako-Rob 80 made by Hako was tested at Waterloo International Terminal in London. The BAROR and AUROR made by Cybernétix were introduced onto the Parisian Metro. These robots are guided by magnets set into the floor. The AUROR weighs 660 kilograms and measures 2000 by 730 by 1080 millimeters. It has a maximum working speed of 0.75 meters per second. Autonomous floor-cleaning machines are also manufactured by the French companies Midi Robots and Robosoft, by Denning Mobile Robotics in Pittsburgh, PA, and by the Canadian company Cyberworks.

Navigation Using a Gyro Compass

In Germany, in addition to the Hako and Siemens-Hefter companies, the Alfred Kärcher company is a leading manufacturer of cleaning robots. Its BR 700 robot, a self-teaching scrub/vacuum machine, navigates without the need for artificial markings by way of a gyroscope, displacement transducers, and ultrasound sensors. Safety fenders and contact strips serve to further protect people and objects.

At the Cologne IRW trade fair in 1997, Siemens-Hefter demonstrated the ST-111, which is equipped with ultrasound sensors, laser scanners, and an optical gyroscope. A special feature is its variable working width.

ACROMATIC, Hako, Germany

BR 700 Robot,
Alfred Kärcher GmbH,
Germany

Future Intelligent Environment Perception Systems

Cleaning robots are usually autonomous systems based on complex navigational techniques. Although French design engineers have been quite successful with their robot-guiding system using floor magnets along meandering tracks, the trend nowadays is towards navigation without artificial reference points.

A modern robot feels its way with sensors, creates a map of its environment, registers any changes in it, recognizes mobile objects, moves out of people's way making a friendly comment as it does so, and entertains them with music.

An extensive sensory analysis makes such things possible: laser scanners and infrared sensors are responsible for navigation, ultrasound sensors for avoiding collisions, and touch-sensitive strips and fenders for protection against direct collision and for stopping in an emergency. Sensors on the upper- and undersides of the robot recognize stairs and hanging objects.

In addition to railway stations and airports, cleaning robots will also be seen more and more frequently in supermarkets and shopping malls in the future. For instance, robots for use indoors have been implemented by the larger hotel chains.

At the moment, the purchase price of an autonomous floor-cleaning machine for the commercial sector is in the region of $28,000, but this price will fall with increasing demands and advancing technological developments.

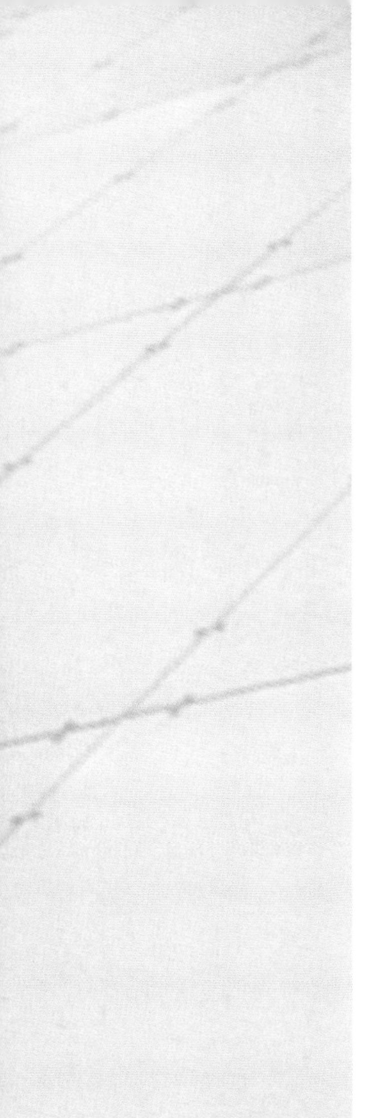

Cleaning Facades

High Tech at Lofty Heights

Glass as a material for building facades is becoming increasingly popular among architects. A glass facade gives an aesthetically pleasing exterior and optimal light conditions inside, but to maintain its beauty and lighting benefits, a glass facade must be cleaned at regular intervals. For the facility managers of large buildings, the expense of a major cleaning every few months is an important cost factor that weighs heavily on the economics of the building.

In the case of complicated glazed surface structures, such as those found in modern airport terminals or railway station halls, additional conditions complicate the issue even further: There are construction obstacles and delicate flooring, and high safety requirements for protecting passengers must be met. The inadequate assessment of functional needs in the planning phase of a building, as well as unsuitable and outdated techniques for accessing heights, often represent insurmountable hurdles for conventional cleaning and maintenance of such large constructions, which sometimes have complex surfaces as well.

The quality level attainable by manual cleaning is influenced by factors such as weather, energy, and the level of motivation of the cleaning personnel. Permanent visible dirt and, in the worst case, irreversible damage to the building are possible consequences.

Manual Cleaning Is Expensive and Dangerous

The accident rate, which is already high for cleaning elevated areas, goes up even further if safety regulations are not followed when accessing difficult areas. In view of rising personnel costs, such areas are often neglected now because of the excessive costs involved. This lowers the value of the building and increases the cost of repair and upkeep.

Facade cleaning exemplifies the advanced mechanization of monotonous and dangerous cleaning tasks. Manual cleaning performed using open- or closed-ladder systems, facade elevator installations, maintenance bridges, and hydraulic working platforms are all part of today's technology.

Trend Towards Automating the Cleaning of Facades

Cleaning facades in particular, with its corresponding high safety and working place quality requirements, offers a huge potential for automation and robotic technologies; after the area of floor-cleaning, mechanization in this area is the most advanced. The basic technologies required for cleaning facades are already in existence and further key components will be available soon.

A questionnaire devised by the Fraunhofer Institute for Manufacturing Engineering and Automation (IPA) in Stuttgart, Germany, was sent to numerous service companies for cleaning buildings and showed a clear result: Manually cleaning the facades of large buildings has become almost impossible from an economic point of view because of the huge glass surfaces and complex facade structures involved. International experience substantiates this: Particularly in the case of expensive buildings with a long life expectancy, facility managers and the initiators of building projects are prepared to invest in automated maintenance and cleaning.

High Safety Level and Improved Performance

The most obvious advantage of automated facade-cleaning systems is seen in the revised safety requirements. With automated systems inaccessible to personnel, the requirements are limited to the general safety of the installation, so workers are no longer endangered.

In addition to relieving human workers from monotonous and potentially dangerous cleaning tasks, automated cleaning systems offer a variety of other advantages:

- Increased cleaning performance with a better and more consistent cleaning quality that is less harmful to the environment.
- Improved value retention of buildings and installations.
- Lower cleaning costs because additional measures such as closing off the building to the public and using extra equipment can be spared.
- The system has other useful functions, such as inspecting parts of the facade using video cameras.

The State of Automatic Facade-Cleaning

The first facade- and glass-cleaning robots were developed and put into operation in Japan in the 1980s. These robots had to cope with problems unfamiliar to other systems, such as maneuvering over joints and being equipped with a sensor to provide quality control of the cleaning process.

The level of automation in current cleaning systems ranges from simple remote-controlled solutions to systems that have completely automated the cleaning process; these systems are already in use in Japan, France, and Germany.

Facade-cleaning robot, Comatec, France

The First Glass-Cleaning Automated Machine at the Leipzig Trade Fair

The world's first fully automated glass facade-cleaning system for domed glass halls works on the glass roof of Leipzig's exhibition hall. It was developed by the Fraunhofer Institute IFF in Magdeburg, Germany. The domed glass roof of the exhibition hall in Leipzig is 250 meters long, 80 meters wide, and 28 meters high at the crown. In order for the exhibition hall to retain its transparency and to be well-lighted, the 25,000-square-meter exterior surface area must be cleaned at regular intervals.

The cleaning robot at the Landmark Tower in Yokohama, Japan, is one example of a robot that has been specifically developed and adapted for the building by the technical departments of construction companies. The cleaning robot operated by the French company Comatec cleans the glass pyramids in front of the Louvre in Paris.

Facade-cleaning robot, Fraunhofer IFF, Germany

Fraunhofer Institut
Magdeburg

How do Cleaning Robots Work?

To date, prototypes of glass- and facade-cleaning robots made in Germany, France, and Japan have been based on two typical concepts of motion: a robot moving strictly along tracks, and a free-moving robot that travels over facade surfaces and is attached to a system from above (from the roof).

These are only two of several possible concepts. The type of building, the size of the area to be cleaned, and the point in time at which the cleaning aspect is included in the building plans are all factors which determine the design of the automated cleaning machine. Generally speaking, specific solutions are developed for large buildings. In the case of small buildings, standard systems tend to be considered.

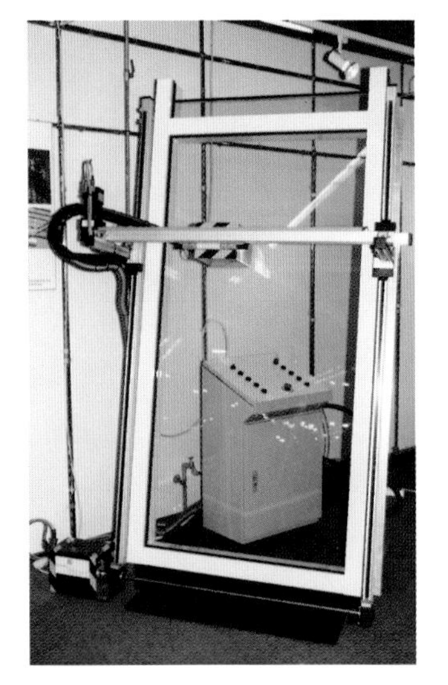

Prototype of a standard facade-cleaning robot, Dornier Technologie and Fraunhofer IPA, Germany

Demands from Practical Use

The following scenario describes how a standard cleaning device is used: A service provider owns a transportable cleaning robot for cleaning buildings. He or she hangs the device in a receiving apparatus on the exterior of the building to be cleaned. The supply aggregate, which stays on the ground (mounted on a trailer), is then connected to the robot via cables and hoses, and the start button is pressed. The robot then travels up and down the invisible guiding tracks integrated into the facade and cleans the facade or windows as programmed.

The requirements of facade-cleaning robots are determined partly by the architect and partly by the service provider or the facility manager. When designing a building, the architect does not want his freedom limited by a cleaning device and is not prepared to accept any aesthetic disadvantages affecting the building's appearance.

The requirements of the service provider are more practical. Above all, he wants a system that is easy to operate, needs little maintenance, is easily transportable, and can be adapted for use on many buildings. He demands a high level of cleanliness without harming the environment and without running the risk of damaging the facade or disrupting business in the building. Low operating costs and an attractive design are also high on the list of requests.

Subsystems of a Cleaning Robot

A facade-cleaning robotic system is made up of various components, each fulfilling individual demands.

The robot's guidance system and its method of attachment to the building's facade must meet high safety requirements and allow the device to move around on the facade. Especially in the case of standard cleaning robots, the guiding elements should blend harmoniously into the architecture of the building and not be distracting. Suitable solutions are currently being developed, particularly in conjunction with building facade manufacturers.

The transport system, with its drive unit and kinematics guiding the cleaning head on the facade or pane of glass, are technologies already in existence which need to be adapted to fit the guiding concept. The transfer of the transport system from one rail track to another can be carried out manually at the base of the building.

The cleaning head is the key part of a cleaning robot and is responsible for the actual cleaning process. Highly effective prototypes already exist in this field, such as the brush cleaning head developed by Dornier Technologie (Uhldingen-Mühlhofen, Bodensee, Germany) which guarantees smear-free window panes.

The washing medium used is completely recycled during the cleaning procedure.

Alternative technologies, such as ultrasound cleaning, work without the macroscopic-mechanical movement of a cleaning tool, with the result that no scratches occur on the facade. There is still a need to develop an automatic cleaning head that has a sensory analysis for recognizing dirt and a quality monitor integrated into it.

In contrast with the facade-cleaning robot prototypes implemented to date, the supply of media (cleaning agents, energy) to a universal standard cleaner designed for service providers should take place from the ground. In the case of a robot for large buildings, the energy is wisely supplied by conductor rails in the tracks and cleaning agents are supplied autonomously by the device itself. Today, it goes without saying that the cleaning agents must be recyclable, and that environmentally friendly cleaning substances must be used.

More Acceptance from Architects and Service Providers

Control of a cleaning robot must be particularly user-friendly so that cleaning enterprises are willing to use automation technology.

Architects and facade-cleaning companies must be encouraged to embrace service robots. Only then can automation be applied to facade-cleaning so that many cleaning processes still carried out manually can be performed automatically. The level of automation must be specifically determined for each case in question; economic and technical feasibility form the basis of each decision.

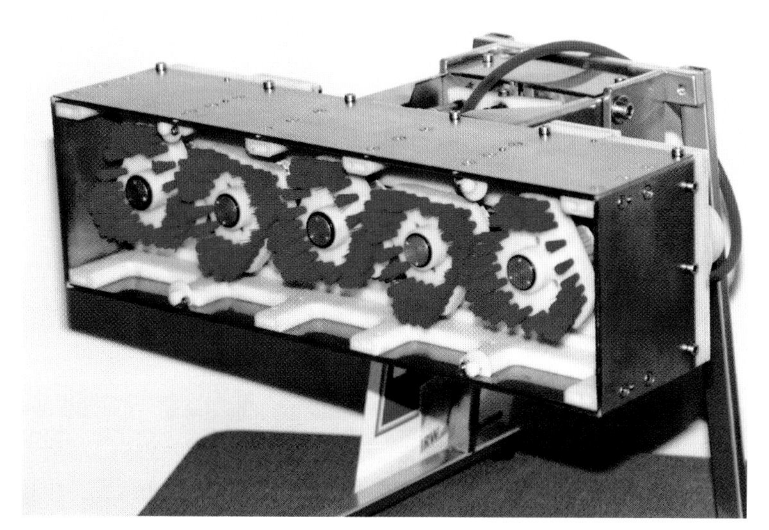

Prototype of a cleaning head, Dornier Technologie, Germany

Aircraft-cleaning robot
Skywash SW33, Putzmeister AG,
Germany

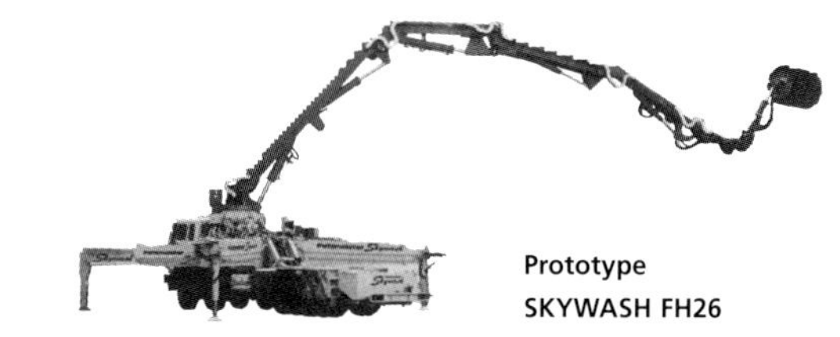

Prototype
SKYWASH FH26

Cleaning Aircraft

Tending Airplanes

All air travel companies conduct bitter price wars with their rivals. Today an airline must use its fleet of airplanes as efficiently as possible in order to remain competitive. This is done mainly by minimizing the length of time airplanes are on the ground and by maximizing the number of passengers.

Grounding of planes is caused partly by maintenance and repairs; for reasons of operational safety, there is little room here for saving money. Service jobs such as catering, refueling, and cleaning, however, also lengthen an airplane's parking time during which it is unable to generate revenue.

Cleanliness is Good for the Airplane

Many airline passengers believe that a clean plane must be a safe plane. An immaculately clean fleet suggests solidarity and a responsible attitude towards technology, comparable with a car enthusiast who always keeps his car spotless and parked in the garage.

Cleaning passenger aircraft, however, does not just serve to maintain an image, it also sustains its material value: Substances settle on the outer skin of a plane during flight and damage the protective coating, and could even lead to corrosion of the aluminum skin.
For this reason, it is a rule of most large airlines that the outer skin of passenger aircraft must be cleaned regularly.

Lufthansa Operates the Largest Service Robots in the World

As is the case with many other cleaning tasks, the cleaning of airplanes is usually still performed manually. A few automatic cleaning systems do exist in Japan and are very much like automatic car-washes. But these systems have some disadvantages: The plane must be towed to the cleaning installations, and only certain types of aircraft may be washed in them.

In Germany, passenger aircraft are also cleaned by hand with a scrubbing brush, with one exception: in Frankfurt, Lufthansa has implemented the two largest mobile robots in the world for this purpose. Together with technical and scientific support from AEG and Dornier, and with the Fraunhofer Institute, IPA, Putzmeister AG in Aichtal, Germany, has developed an aircraft-cleaning robot called Skywash.

Cleaning Time Reduced by 60%

Skywash is highly flexible and effective; by using two Skywash robots, for example, the ground time allocated for cleaning a Boeing 747-400 jumbo jet could be reduced from 9 hours to 3.5 hours. Over this period of time, the washing brush of Skywash, the world's largest service robot, travels a distance of approximately 3.8 kilometers and covers a surface area of around 2,400 square meters, which is about 85% of the entire surface area of the plane. Even the exterior of the engines of this giant bird gets a wash in the cleaning procedure.

off-line programming

Off-line programming may be considered as the process by which robot programs are developed, either partially or completely, without requiring the use of the robot itself.
This includes generating point coordinate data, function data, and cycle logic. (Handbook of Industrial Robotics, 2nd Edition, 1999 John Wiley & Sons)

Precision Despite the Heaviest of Loads

The essential vehicle of this large mobile robot is composed of a reinforced chassis made by Daimler Chrysler and equipped with a 380 horsepower diesel engine with reduced exhaust emissions in accordance with EURO2. Four supporting legs and a manipulator arm 33 meters in length and weighing 22 tons are mounted on the chassis. The robot has eleven programmable joints and can drive a payload of 500 kilograms securely and accurately. The redundant kinematics consist of five main joints, five joints for the robot's hand and, for programming reasons, a further joint for controlling the turning circle of the rotating washing brush.

The autonomous robot transports with it all the subsystems it requires for operation. Specially developed robot control components, an onboard computer that functions as an interactive man-machine interface for communicating with the operator, a highly specialized sensory analysis, and two tanks containing cleaning agents are all integrated as subsystems into the vehicle.

Control Data from a CD-ROM

When preparing Skywash to clean a plane, the operator inserts a CD-ROM into the onboard computer. On this CD-ROM is data concerning the geometrical features of the aircraft to be cleaned and programs that describe the robot's motion path.

In the case of large mobile manipulators, it is no longer possible to create robot programs with conventional teach programming due to the fact that the starting position is not always the same. For this reason, a new type of off-line programming system has been developed by Fraunhofer IPA for creating the motion programs: In a simulation environment, the program devises energy-optimized paths of travel for the robot, taking into consideration the overall geometrical situation and critical collision areas.

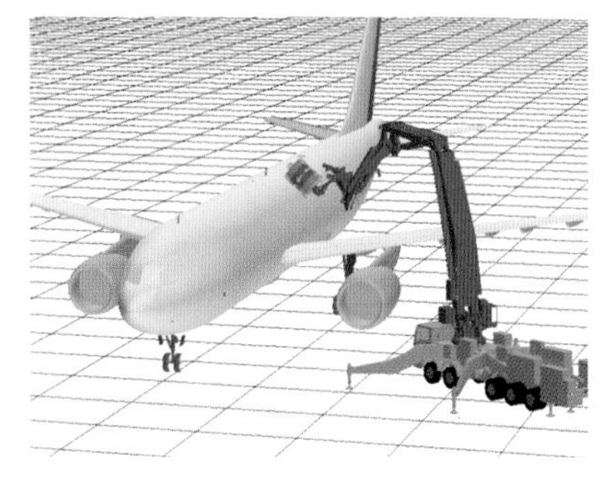

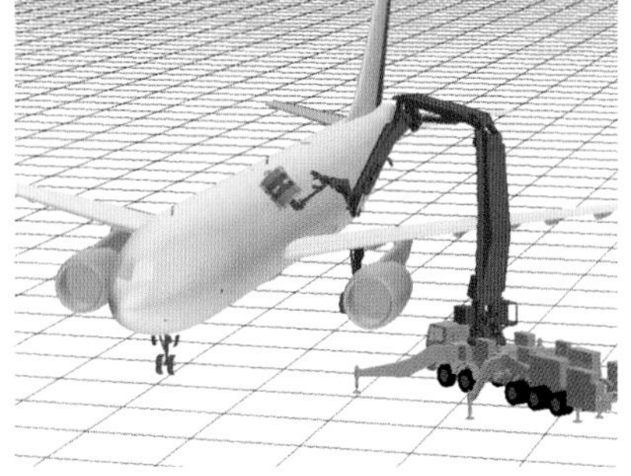

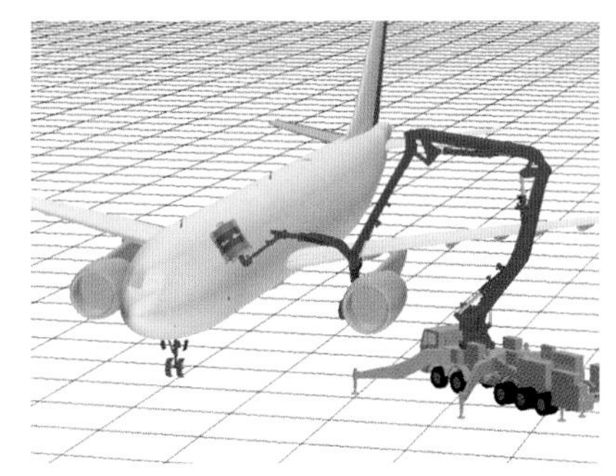

Automatic Positioning Using Distance Cameras

In order to position the mobile robot accurately next to an aircraft outside the hanger without any markings on the plane or on the ground, Skywash has been equipped with a 3-D distance camera (EBK) developed by Dornier. During the approach phase, this laser sensor guides the aircraft into the correct position. The supporting legs of the chassis are activated, and the exact position and orientation of the plane relative to the manipulator is determined from a further measurement made by the distance camera.

The motion path programs loaded onto the onboard computer refer to Skywash's theoretically ideal positioning point but, by using the data obtained by the 3-D distance camera, the reference can now be transferred to the actual positioning point. Once this operation has been completed, the programs are loaded onto the AEG robot control and the actual cleaning process can begin.

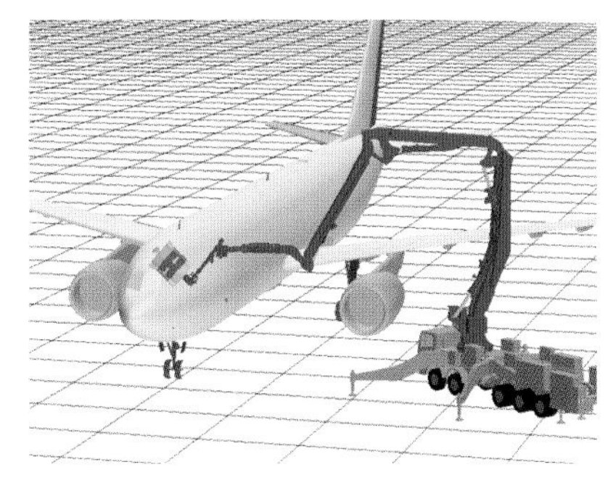

Skywash SW33,
IGRIP simulation

Sensor-Aided Guidance

The arm segment is unfolded first so that the robot's "tool center point" is in the starting window of the first wash program; the arm then guides the rotating cleaning brush over the outer skin of the aircraft.

The washing brush also functions as a sensor which measures the distance between the rotational axis of the washing brush and the surface of the plane. This measurement is derived from the torsion moments, which vary according to the immersion level of the brush. This measurement value is used as an input for a control algorithm that guides the joints of the robot's hand in an optimal way.

Big Savings Make for a Short Amortization Time

Once Skywash has finished at the first positioning point, the robotic arm is then folded up, the stabilizing system with its four supporting legs is retracted, and the mobile robot travels to the next position. Jumbo jets, such as the Boeing 747-400, require four positioning points; smaller planes such as the Airbus A321 need only two.

The grounding costs of a passenger aircraft are in the neighborhood of nearly $3000 per hour. Thanks to Skywash, over $14,000 can be saved per washing process. The purchase price of about $2.8 million per device is amortized within just a few years.

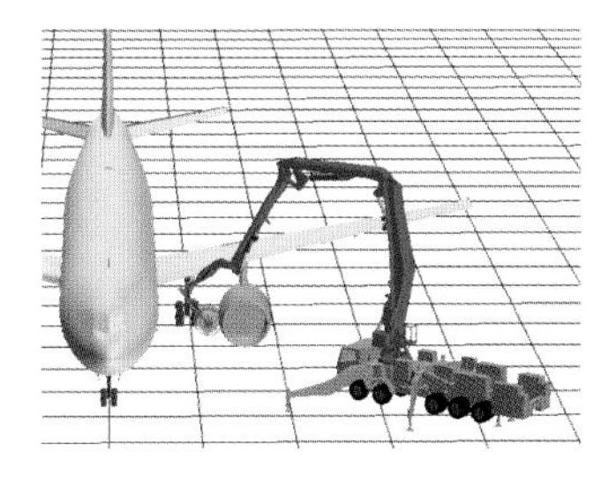

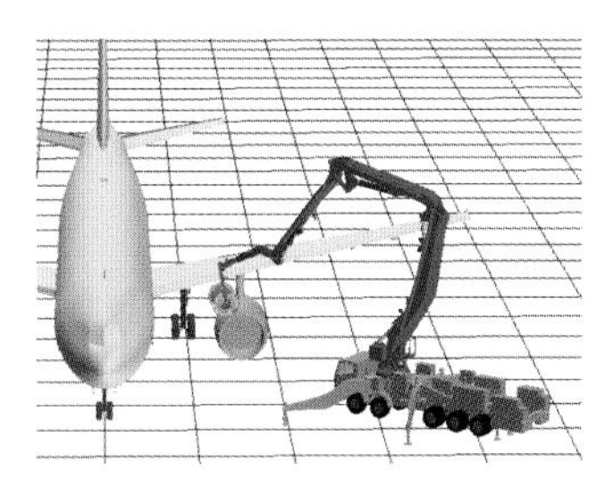

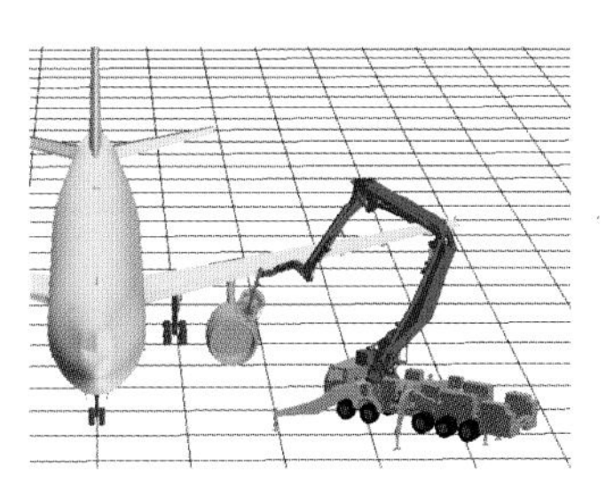

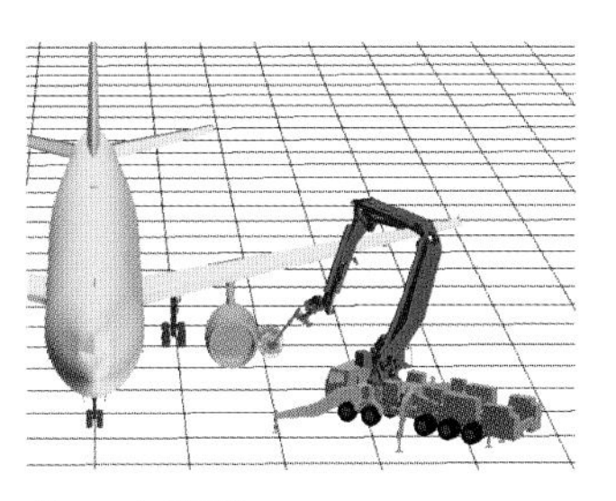

Skywash SW33,
IGRIP simulation

Cleaning Recreational Boats

Robot Intelligence Replaces Environmental Toxins

Nature stipulates certain underwater conditions which don't always suit humans or technology. All materials become colonized by various organisms within a short period of time, a fact that impairs their functionality and durability.

For this reason, surfaces in contact with water in harbors and offshore installations, fishnet cages in aqua-culture, and ships' hulls or markers such as buoys are usually coated with anti-fouling chemicals and special underwater paint.

Toxic substances are given off slowly and continuously by the coating to prevent regrowth. The biocides generally used are intended to impede colonization by organisms and are either organic zinc or cupric compounds or substances called co-toxicants, such as organic compounds containing nitrogen.

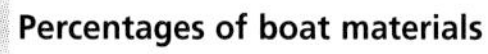

Percentages of boat materials

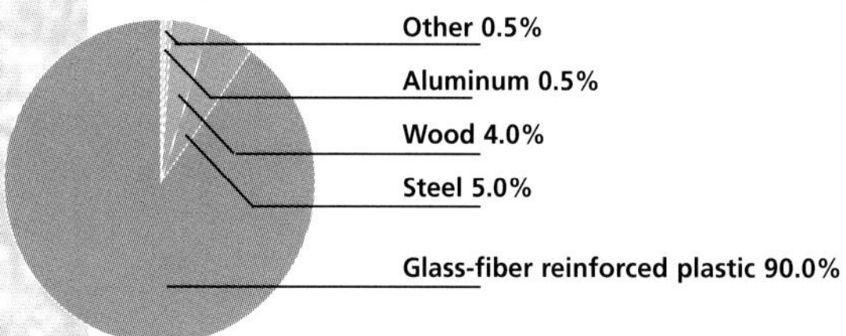

Amounts of waste arising from boat-cleaning processes, Amount of waste generated from the cleaning of 100 boats

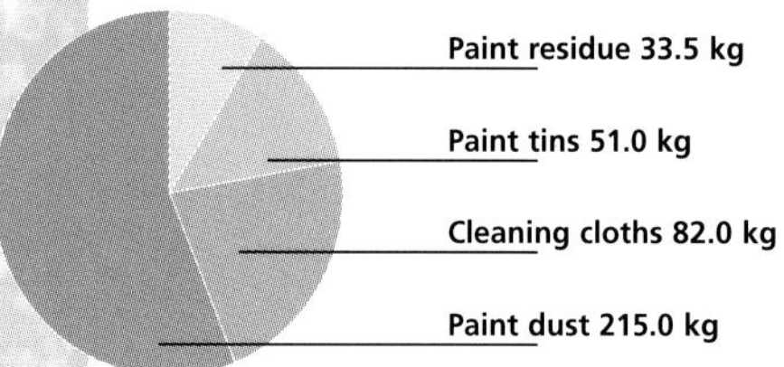

The potential danger to the environment is well known. It has already been proved that pollutants with a long half-life accumulate in the sediment, and that this affects sea life such as oysters and sea snails, which in turn form part of the human food chain.

Biocides Threaten the Marine Environment

In Germany, boats that stay in the water for the entire season because of their weight are still usually coated with an anti-fouling paint containing biocides. Only smaller boats, such as dinghies, are removed from the water after use and therefore do not require an anti-fouling coating.

Today German federal nature protection laws do not permit the use of underwater coatings containing biocides for leisure motorboats or sailing boats. They grant people recreational rights in the countryside but oblige them at the same time to treat the environment's resources carefully. These laws will soon be modified in order to permit any kind of marine pollution. It is imperative that the burden on the marine environment caused by recreational boats be kept to a minimum.

Alternatives to Toxic Chemicals

In Sweden, the use of anti-fouling coatings containing biocides on boat hulls is already forbidden in inland waters. The European Union is presently considering whether to follow the Swedish example.

What alternatives to growth prevention exist if no toxic underwater coatings may be used

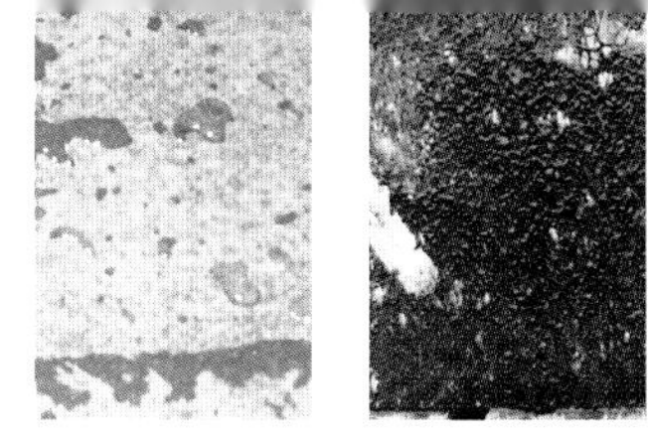

Cleaned and colonized boat hulls

Growth after approximately 4 weeks at the coast: mechanical cleaning no longer possible

Growth after approximately 3 weeks at the coast: mechanical cleaning still possible

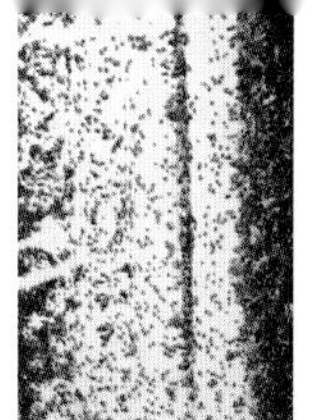

Plate following mechanical cleaning

any longer? No owner would like a slimy coat of algae on his boat, increasing the frictional resistance in water and spoiling its appearance.

Most inland harbors and yacht marinas are equipped with cleaning bays and high-pressure wash systems, onto which recreational boats can be slipped or hoisted at the end of the season for cleaning before being laid up for the winter.

At a slip installation, the boat is hoisted onto a submerged cradle and pulled out of the water up a sloping surface. Larger harbors have stationary or mobile cranes that hoist the boat out of the water and place it onto a trailer. Usually, the boat is then transported from the crane site to a special wash bay. There, the boat is cleaned with a hand-held high-pressure cleaning device.

Anti-fouling paints containing biocides are relatively unstable by nature. The high-pressure cleaning process abrades the coating, and particles of this coating must not get into the water supply. For this reason, these facilities must be equipped with a closed-circuit water system.

Crane site and washing area

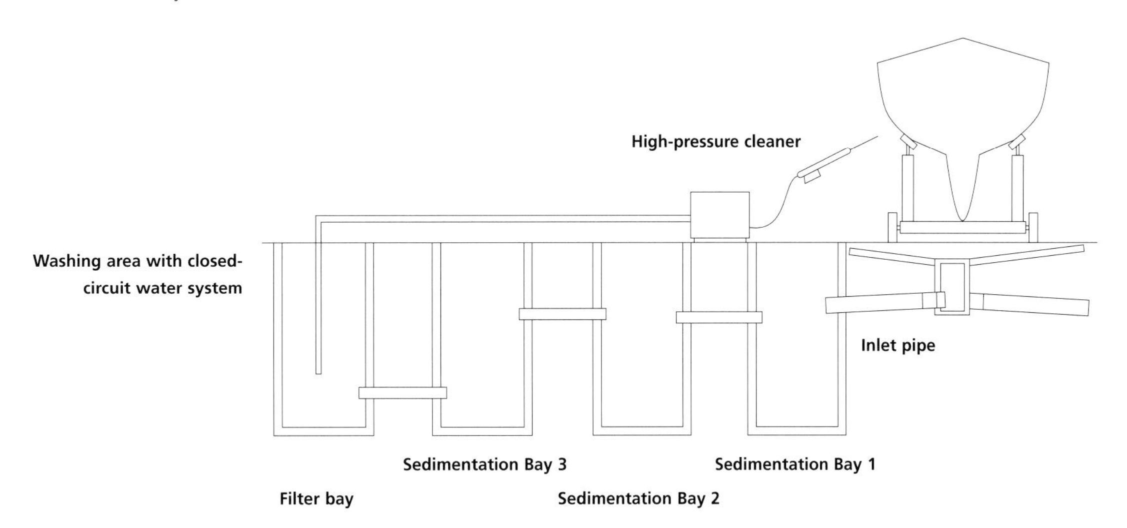

Washing area with closed-circuit water system

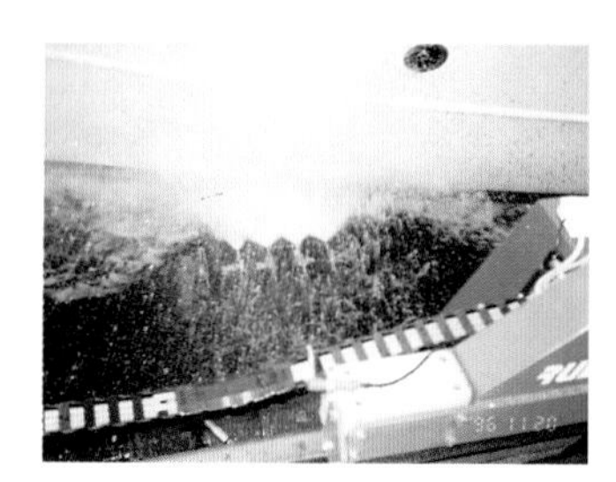

Automatic boat cleaning system,
RULE Battvaetten AB, Sweden

Automatic boat cleaning system,
RULE Battvaetten AB, Sweden

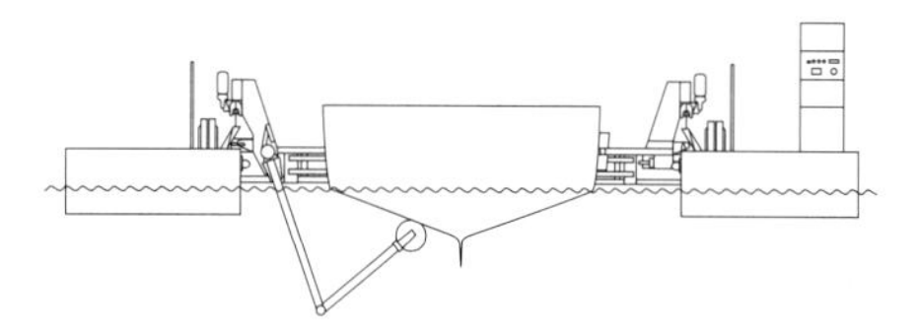

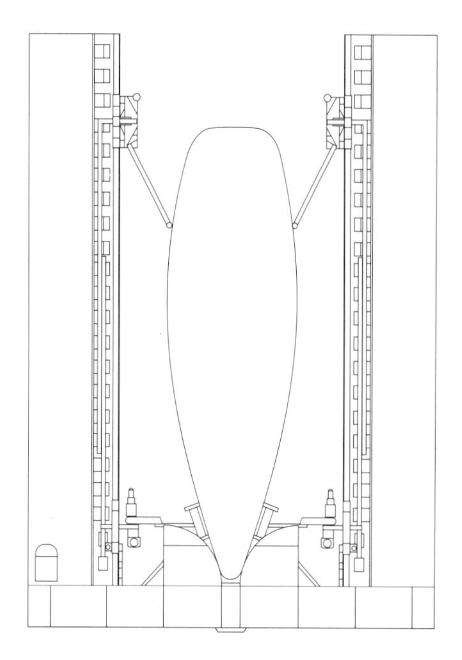

KBK Boatcleaner, Sweden

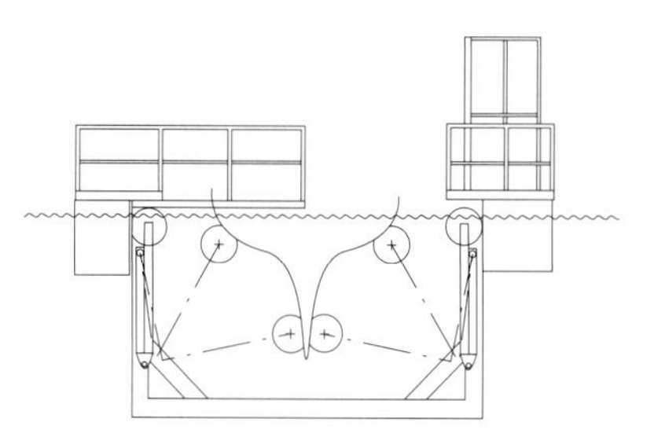

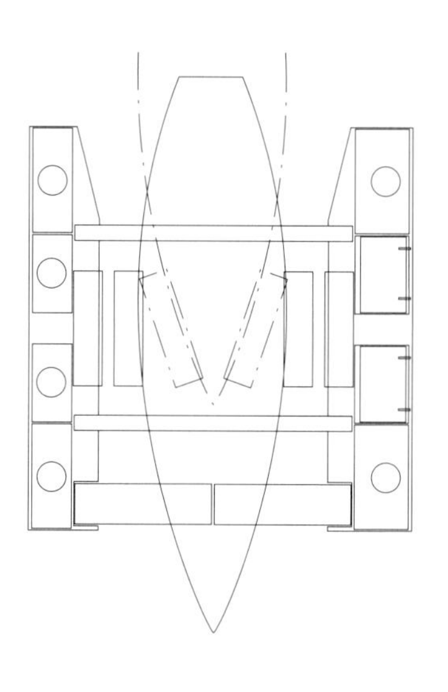

Sweden Already Manufactures Systems

Swedish laws have led to the development of a mobile wash system for motorboats, the Stark Boat Washer. This system, which was introduced by the Starkmatic company in 1994, is a transportable, floating cleaning system for small motorboats up to three meters wide. It can be transported on a trailer and is quickly erected and dismantled.

Using a winch, the boat to be cleaned is pulled up over two pairs of revolving brushes arranged in a V-shape. The boat is lifted up by the brushes during the process; the pressure between the brushes and the hull of the boat is dependent on the boat's weight and the surface area resting on them. Due to the rigid arrangement of the pairs of brushes, only motorboats with a curved frame construction can be cleaned.

Anti-Fouling on Demand

An important demand in the mechanical process of cleaning boat hulls is that biocide-free chemicals be used so that no hazardous substances are released and no contamination of the seawater occurs. When planning a cleaning system, therefore, water and disposal regulations must be observed. As with other innovative lines of development, such as using the principles of electromagnetic fields or coatings manufactured using bio-engineering techniques, design engineers are concentrating on alternative protection methods and, in particular, on robot-assisted systems.

Robotic systems are capable of removing growth from the non-toxic coated surfaces of boats with minimal abrasion in salt and fresh water. This is known as anti-fouling on demand.

Brushes and Seawater Under High Pressure

Because of its complex shape, the underwater section of sailing boats represents a particular hurdle to automatic cleaning systems. Several patents exist, but to date only two systems for cleaning recreational boats are available as prototypes.

The system made by the Swedish company, Rule, consists of one or two movable robotic arms with four high-pressure jets in their extremities. The boat is maneuvered into a floating box and clamped to a centering device. Sitting on the floating pontoons, the robotic arm moves underwater on supporting wheels and, moving from top to bottom, travels in rows along the contours of the boat.

For the high-pressure jets, seawater is used. The 0.5 meter wide cleaning head has a speed of 250 millimeters per second. At this rate, between two and three boats, each up to 4 meters in width and 16 meters in length, can be cleaned per hour.

Kim Koch is the inventor of a second system constructed in Denmark. This works in the same way as an automatic car wash. Revolving brushes are attached to manipulators on the left- and right-hand sides of the boat. In this instance, the boat is also cleaned in the water. A problem not yet solved is how to clean the all-important structures on the hull of a sailing boat, such as the transition of the keel to the hull, propeller, and rudder.

Research for the Ministry of the Environment

At the Fraunhofer Institute, IPA, in Stuttgart, Germany, work has been done on a floating cleaning system for recreational boats within the scope of a research project funded by the Ministry of the Environment. The boat to be cleaned is maneuvered into place and is then lifted out of the water. The hoisting mechanism shifts the surface of the supporting belts automatically; the boat does not need to be set down in the interim.

Cleaning Is Carried Out Using High-Pressure Jets

Once the boat has been hoisted up, strips bearing several high-pressure jets are brought alongside. They are arranged in such a way that the complete cross section of the boat can be cleaned in one process. The entire cleaning process requires that the strips be moved along the length of the boat just once. The cleaning procedure is gentle on the environment due to the fact that the residue from the hull and the cleaning water are collected and retained. The waste water can be processed and used again as washing water.

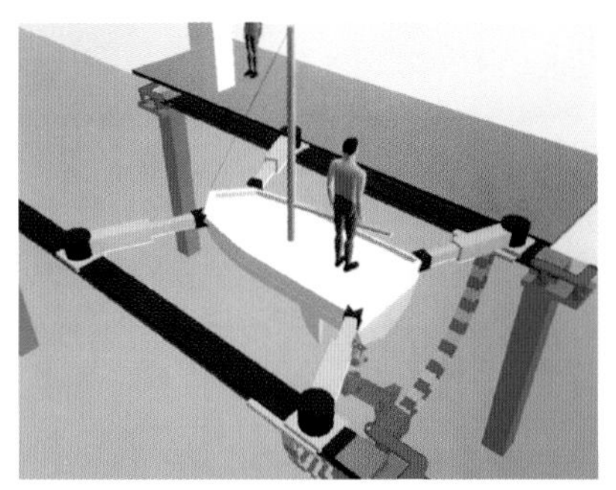

Automatic boat cleaning system - different stages, research project Fraunhofer IPA, Germany

Environmentally Friendly Boat Cleaning Without Human Assistance

The system is suitable for nearly all types of sport and recreational boats. It can be as partially or fully automated as desired. One variation can even be installed on dry land. This is particularly useful when a washing bay with a crane is available that can hoist sailing boats with upstanding masts. The lifting mechanism is then no longer required.

The skipper of the boat is able to operate the system alone; no additional personnel are required. The cleaning process is fast and gentle on the surface; wear or damage to the surface does not occur, not even when removing tenacious growth. Areas that are difficult to reach, as well as delicate hull structures, can also be easily cleaned without damaging the boat.

The entire research project has laid the groundwork for the development of similar cleaning systems for ferries.

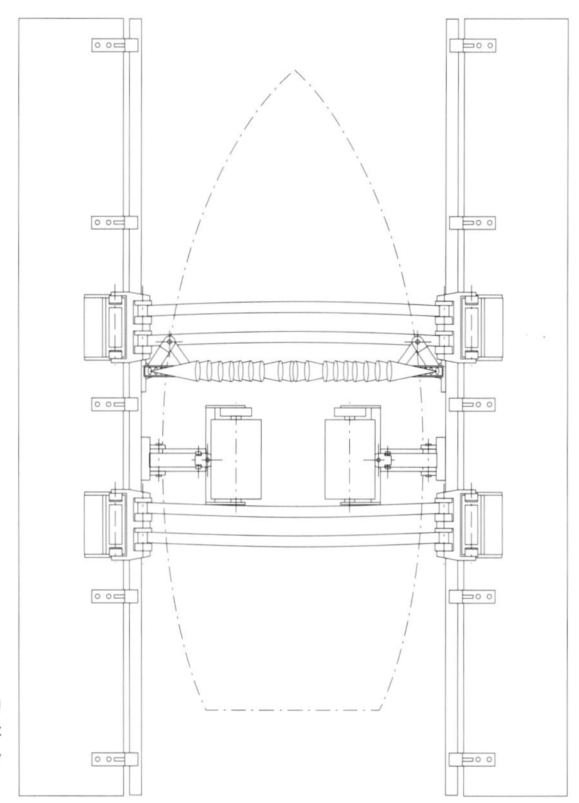

Automatic boat cleaning system, research project Fraunhofer IPA, Germany

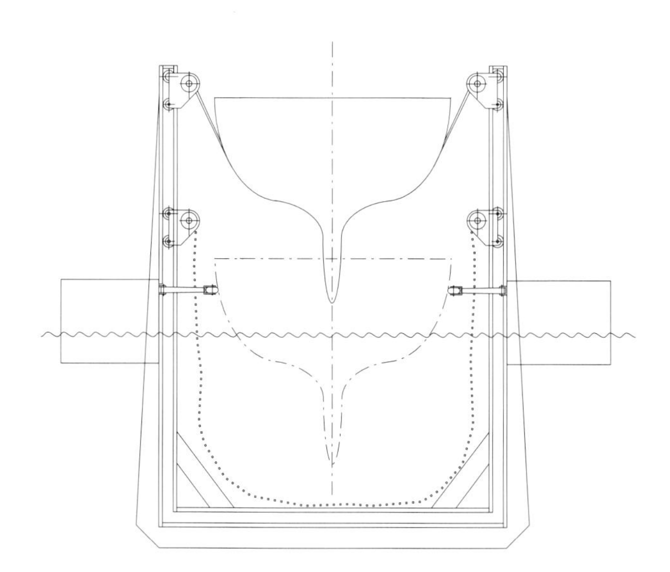

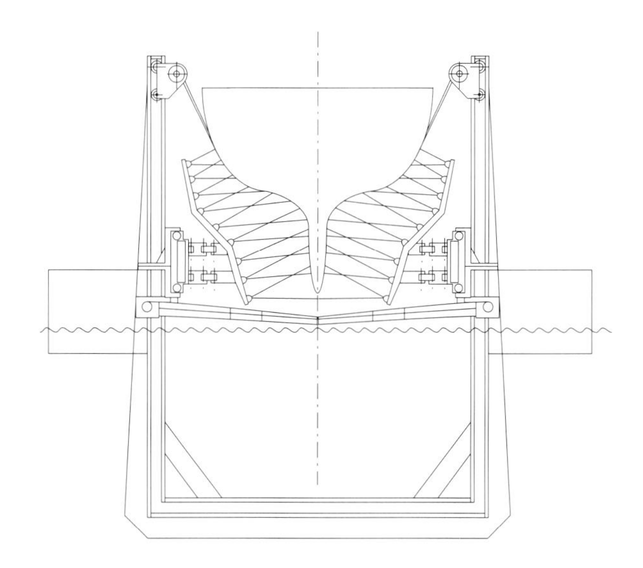

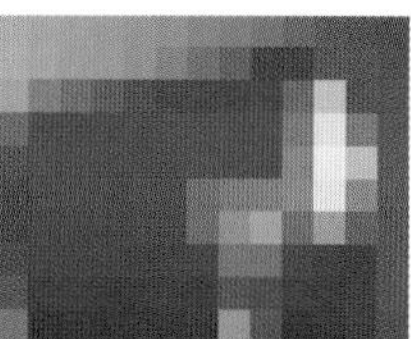

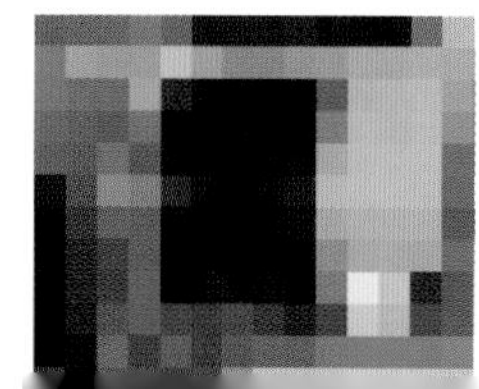

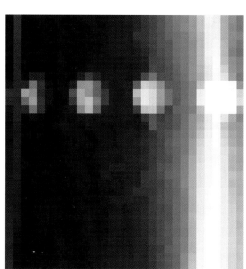

Office

Prototype David,
FAW Ulm, Germany

Automating Office Robots

Mobile service robots are also conquering the office world. They collect and distribute mail, deliver office materials, and dispose of trash. To perform these tasks, these robotic systems must navigate within a typical office environment and plan and execute their tasks without human assistance.

The job is more complex than one may think. An office is actually a highly dynamic and constantly changing environment that demands a great level of flexibility from a service robot. A forgotten swivel chair in the middle of a room or a closed elevator door, for example, forces the robot to make fast and accurate decisions.

Good Sensors Are Necessary

Service robots for performing office tasks are the focus of much development at the FAW Research Institute in Ulm, Germany. David, a research prototype, was created here.

David is based on Nomad 200, an experimentation platform manufactured by the California company Nomadic Technologies. Nomad 200 moves using three steerable drive wheels. With the standard configuration, the mobile robot has a ring equipped with 16 infrared sensors and another ring with 16 ultrasound sensors for environment perception and collision avoidance; two other rings are arranged with tactile sensors.

As an additional navigation sensor, David wears a two-dimensional laser scanner on his head made by the German company Sick. Using this scanner, a depth profile of the robot's environment can be recorded within an arc of 180° with an angle resolution of ±0.5° and a distance resolution of ±3 centimeters.

A color video camera helps David recognize special features and objects around him. Data processing and control tasks are executed by an onboard computer with a 486 processor and 16 megabytes of RAM. Several processes for focused and responsive planning have been implemented for navigation and steering. Processes for determining David's actual position have also been included.

David Notices Full Wastepaper Baskets

A model-based object-recognition system backs up image processing: It extracts specific objects from the video images and determines their exact position in the room. The central software "task-planning" components plan elementary operational sequences, such as moving, turning right or left, and stopping.

One of David's functions is collecting wastepaper baskets in a typical office using a simple but effective forklift mechanism. David performs this task in several steps:

Once the mobile robot has received the command to pick up a wastepaper basket from a specific office, it develops its plan of action. In the first step, David tries to reach the point where he thinks the wastepaper basket is located. Along the way, he uses his collision-avoidance strategies and maneuvers around unexpected obstacles. Once David has reached the place where the wastepaper basket should be, he takes a picture with the video camera. The data processing function module then attempts to extract the feature wastepaper basket from this image. The exact position of the wastepaper basket can be calculated because the position and orientation of the camera are known. David then picks up the wastepaper basket and transports it to the appropriate place.

Ethernet antenna
Emergency shutdown
Console
Console support
Ultrasound sensors
Storage space
Storage space door
IR sensors
Laser scanner
Door
Base
IR sensors
Wheels

Mobile platform B21,
Real World Interface, USA

Mailman with an Internet Address

At the Institute for Robotics at the Swiss Federal Institute of Technology (ETH) in Zürich, Switzerland, the technological and research platform MoPS (Mobile Post-distribution System) has been developed. This robot transports mail between a central point on the ground floor and separate offices within the building.

It takes the container with the mail, which has been manually sorted for the various departments, and delivers it to the five administrative sections on the different floors. MoPS collects any containers with outgoing mail at the same time and deposits them at the central post office at the end of its trip.

The trip is pre-planned but may be altered by an authorized person via the Internet, if necessary. The robot can visit two departments per journey. To change floors, it takes the service elevator. Communication with the elevator and the automatic door-opening system is done via infrared signal.

If Nothing Else Works, It Sends an E-mail

The mail robot works in an unstructured environment. It recognizes obstacles and people, points this out by making a comment, weaves around them if necessary, and goes about its business. Occasionally when MoPS is unable to go any further because, say, the elevator doesn't arrive, it lets the appropriate person know by sending an e-mail to their pager.

A touch-screen on the robot acts as a man-machine interface. Communication with MoPS is also possible via the Internet. MoPS is 60 centimeters wide, 100 centimeters long and 140 centimeters high; it weighs approximately 90 kilograms and can carry a payload of 50 kilograms. A passive sliding wheel and two independent electrically powered wheels give it the required maneuverability.

MoPS adapts its speed according to the environment: If it does not encounter many obstacles or people and is in a wide corridor, it moves quickly; otherwise it slows down. Its average speed is 0.5 meters per second. The batteries allow an operating time of around four hours. MoPS can travel independently to a charging station to change batteries if needed.

A manipulator integrated into the mobile platform can pick up and handle two commercial plastic baskets containing mail up to B4 size (353 by 250 millimeters). The containers can weigh up to 15 kilograms and are on height-adjustable tracks. The gripper attached to the manipulator pushes the containers into a mailbox or picks them up from there; thus the mailboxes have been constructed as inserts in cupboards in the corridors. They can hold up to three containers and are closed with roller blinds that MoPS opens and closes using an infrared control.

Millimeter-Precise Positioning with IR

Currently, MoPS uses a power PC as an onboard multiprocessing computer. If it requires extra computing power for image processing, for example, additional processor cards can be used. The object-orientated real-time operating system XOberon is implemented. XOberon was developed by the Institute for Robotics at the Swiss Federal Institute of Technology (ETH) in Zürich, Switzerland. It is based on the object-oriented standard language Oberon, a further development of the programming language Pascal.

Research platform MoPS, ETH Zürich, Schwitzerland

To find its way around, MoPS uses series of sensors. The most important sensors are two laser scanners at the front and rear of the vehicle. They enable the robot to make a rough estimate of its position, to navigate safely, and to recognize obstacles in its surroundings. At a distance of 10 meters, it has an angle resolution of ±0.5° and an accuracy of ±10 millimeters.

To load and unload mail boxes and to change its batteries, the robot must be able to navigate with an accuracy of ±1 millimeter, due to mechanical tolerances having been kept very low. To attain this degree of positional accuracy, MoPS is equipped with two further infrared triangulation sensors for fine positioning. These sensors have a range of only 50 centimeters, but are so precise that they can make fine corrections to MoPS's position at strategically important points. A contact strip made up of six segments functions as a tactile safety sensor and stops the vehicle in case of a collision.

MoPS, ETH Zürich, Schwitzerland

Robots Will Be Capable of Learning in the Future

MoPS should soon be equipped with extended image processing capabilities. The first image-sensor prototype to be developed can recognize signs on doors; another sensor system will be able to tell the difference between humans and inanimate objects. To achieve this, camera images will be interpreted together with information from infrared sensors, laser scanners, and noise sensors.

Because mobile robots must function in constantly changing environments, the Situation-Based Behavior Selector (SBBS) has been developed for MoPS. This navigational method, which functions in the same way as laser scanner-based position determination, is also based on a map of the surroundings that has been memorized in the form of a "generally plotted graph" and contains strategically important information about the environment. The edges of the graph define the virtual path segments that MoPS should move along.

Among other things, this graph contains individually memorized information about the robot's basic and reflex behavior for each part of its course. The reflex behavior takes over in exceptional circumstances, such as when an obstacle is recognized or when the elevator does not arrive. The SBBS system decides how the robot should conduct itself on the separate segments of its path. At present, the ETH in Zürich is working on the autonomous improvement of MoPS's individual behavioral reactions using an evaluation function (reinforced learning).

MoPS collecting mail

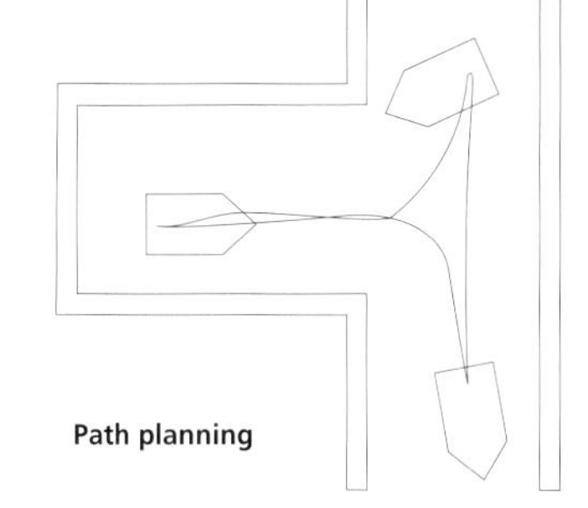

Path planning

Rhino Gets To Know Its Environment

The mobile robot Rhino should be able to move around completely autonomously in dynamic environments and interact with people to carry out its duties. It has been the main research topic for the past few years of the project group Artificial Intelligence (AI) at the Institute for Information Technology III at Bonn University. The AI group is directed by Professor Armin B. Cremers.

Rhino is based on the mobile platform B21 made by the U.S. manufacturer Real World Interface in Jaffrey, New Hampshire. The B21 is equipped with 56 infrared sensors that react to obstacles in their immediate vicinity and a further 56 tactile sensors that react to touch. Two two-dimensional laser scanners, made by the German company Sick, and 24 ultrasound sensors supply distance measurements that are used to generate a digital map of the surroundings. A swiveling stereo color camera completes the mobile robot's sensory equipment.

The aim of the research scientists in Bonn was to develop methods and algorithms which would permit Rhino to explore his surroundings and to make a map of the environment. The system that they developed combines data obtained from the various distance sensors, based on probability methods, and can generate metric maps.

Topological graphs are extracted from these maps, giving a symbolic representation of the environment in compact form. These graphs form the basis for navigational plans for the robot, guiding it on the best path between two points.

Working Safely with People

Dynamic Window Approach is a process developed to allow Rhino to achieve the highest possible level of safety when avoiding obstacles. This process utilizes data integrated from up to six different types of sensors. As a result, Rhino is able to move around quickly and safely within his surroundings.

The "position probability grids" process was developed to determine the mobile robot's global position within its environment. By checking the concurrence of the measured sensor data with a model of the environment, a probability distribution of the robot's position can be approximated using a three-dimensional grid.

The stereo image analysis gives estimates of distances from edges, such as door frames and tables, which cannot be determined accurately by ultrasonic sensors due to the poor reflective properties of corners. It is also used for extracting features, such as objects to be picked up from the floor.

In the summer of 1997, Rhino was ready to be presented to a broad spectrum of the public as he interactively guided visitors around an exhibition held in the German Museum in Bonn. In just six days, the robot accompanied museum visitors to the various exhibits and described these exhibits over 2,000 times. Using its multimedia interface, Rhino also guided more than 600 Internet surfers from the USA, Canada, and Japan, as well as surfers from numerous European countries, around the museum.

MoPS, David, and Rhino are primarily intended to be learning and working models, with whose help the complex functions of service robots can be developed and tested during contact with people.

**Mobile robot Rhino,
Bonn University, Germany**

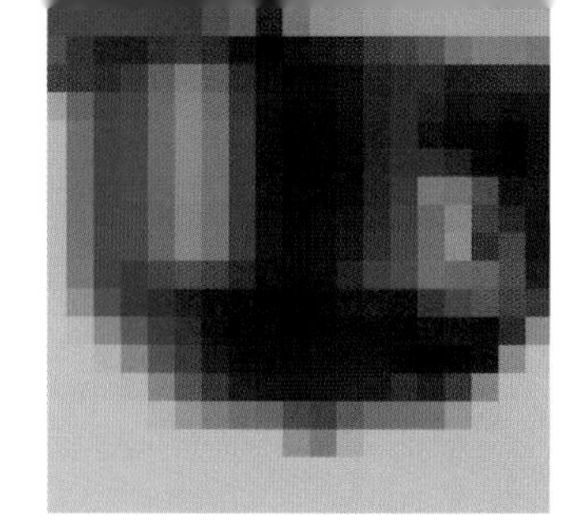

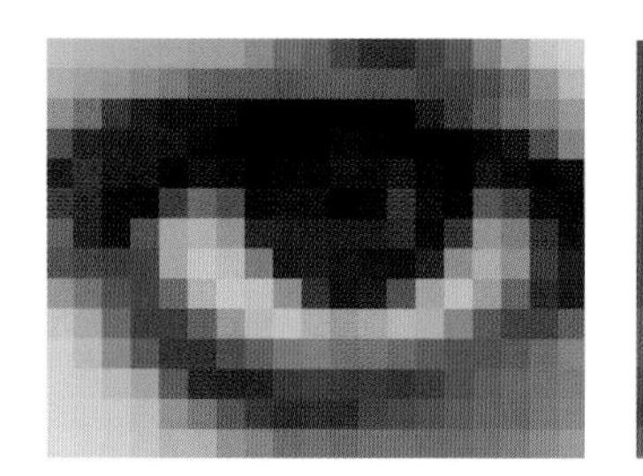
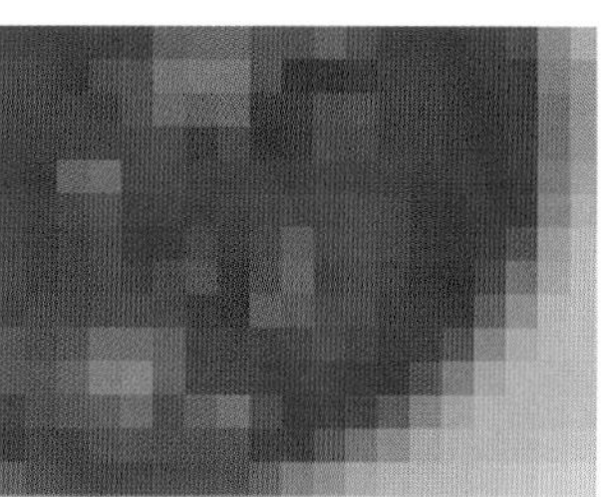

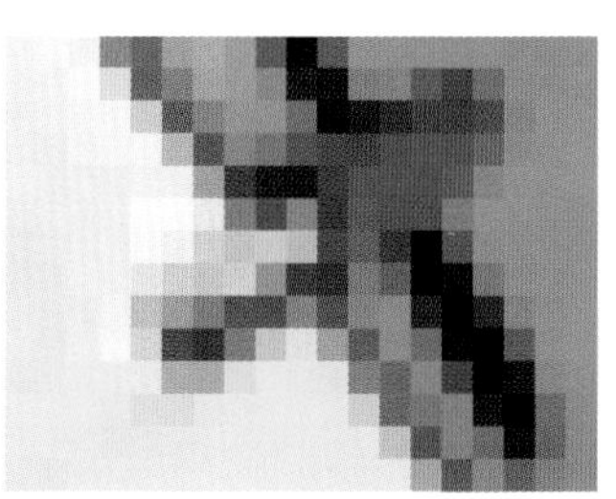

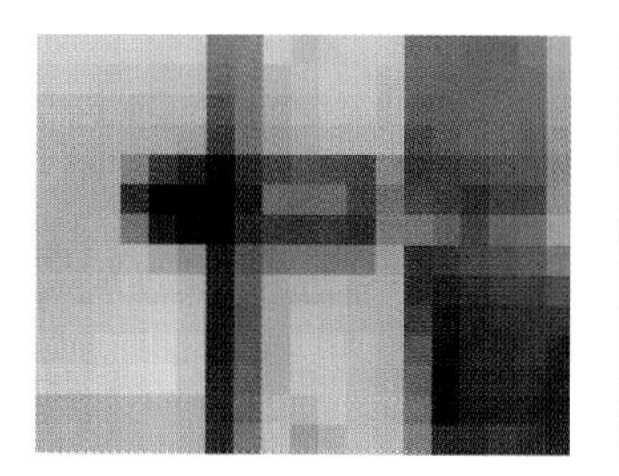

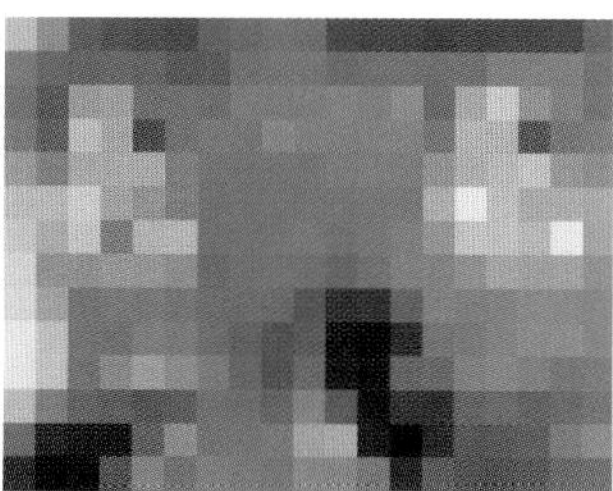

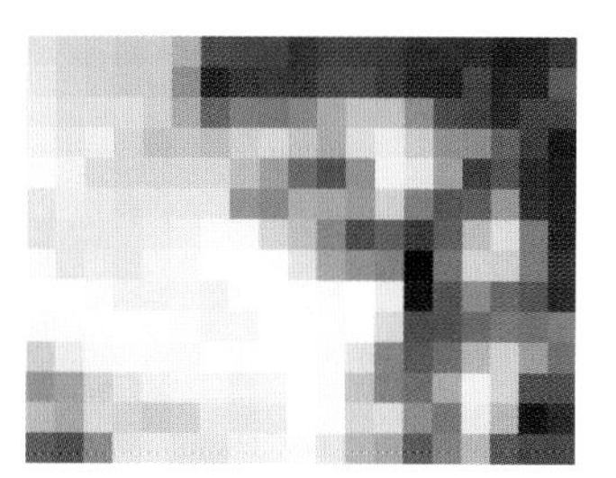

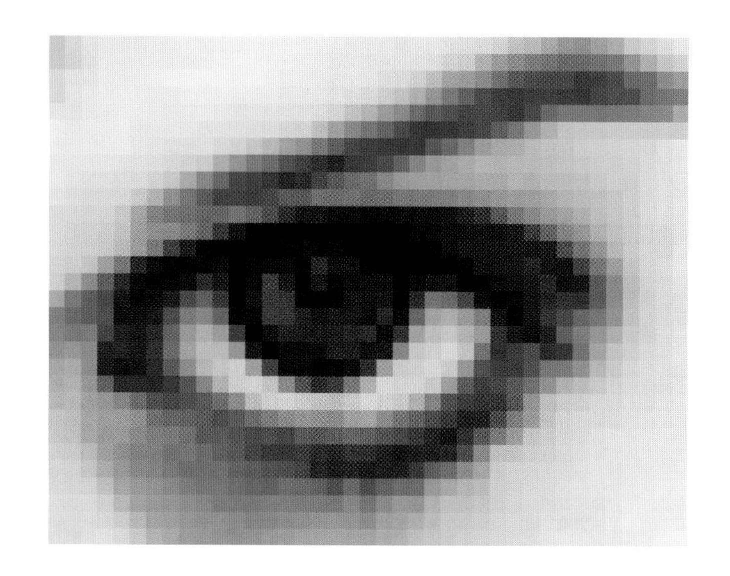

Surveillance

Security With an Overall View

Surveillance cameras can be found in every museum today to ensure that art objects will still be around for the next generation. Cameras, however, can neither follow a thief nor smell a burning cigarette.

It is impossible for human security personnel to be everywhere at the same time, and they are sometimes overworked and often face dangerous situations. If, for example, a lethal concentration of carbon monoxide in the air goes undetected, the guard's life would be in danger.

The surveillance of buildings and installations is an almost ideal application for service robots because they can bring their superhuman recognition skills fully into play. Security investments are almost always worth the money.

Admittedly, George Orwell's 1984 is inevitably brought to mind when providing protection for people and priceless valuables from criminal attack using surveillance electronics. Their benefits and abuses are difficult to separate in such cases.

Assistance for Factory Security Officers

Vigiland is a mobile robot that performs reconnaissance and surveillance tasks on open ground. It supports the existing operational security personnel on the grounds of airports, industrial installations, military establishments, and nuclear power plants. With additional equipment, its area of application can be extended to firefighting and manipulating hazardous materials.

The four-wheeled cross-country vehicle is based on an already existing chassis developed in 1992 by the French company Camiva, a subsidiary of Renault. A 75-horsepower diesel engine powers all four wheels via an automatic transmission. Steering is either servo-assisted or electrical. Without a load, the vehicle weighs 1,900 kilograms and it can transport a payload of 1,100 kilograms. The cross-country vehicle can cross ditches up to half a meter wide and can climb slopes with an incline of up to 45°.

Cybernétix, a French manufacturer of autonomous mobile platforms, has modified this vehicle so that it can be driven like a normal

CyberGuard K2A, Cybermotion, USA

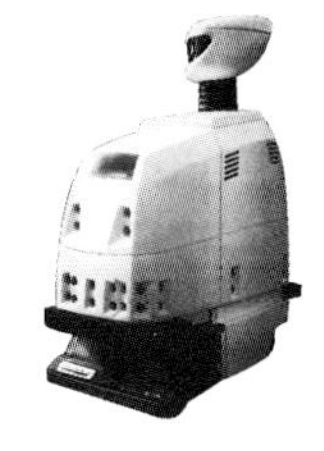

Surveillance robot made by Midi-Robots, France

Inspection robot for pipelines, Japan

Track-guided surveillance robot, Plant Patrol Robot, Japan

car, be remote-controlled, or even function fully automatically; in the latter case, navigation takes place using ground markings. In remote-control and fully automatic mode, the maximum velocity decreases from 120 to 20 kilometers per hour.

CCD cameras and microphones are standard for this reconnaissance vehicle. For firefighting, the robot is equipped with either a water or foam cannon. A robotic arm, also developed by Cybernétix, can be fitted onto the vehicle to allow it to handle hazardous materials.

Vigiland, Cybernétix, France

Mobile Security Guard

Mobile robots for reconnaissance and surveillance of buildings have been developed by the U.S. company Cybermotion since 1984. Their latest model is called CyberGuard 3. The two most important components of this mobile security guard are the navigation system and the sensor system.

The navigation system is based on a digital map of a building's corridors and rooms stored in the robot's onboard computer. Four powered wheels move the system; steering movements are created by varying the speeds of the neighboring drive wheels.

Different sensors assist navigation and record information about the surroundings; this information is then analyzed according to the surveillance task. Ultrasound sensors recognize unexpected obstacles in the robot's path of travel and the robot makes a corrective maneuver to avoid them. CyberGuard 3 can even call the elevator by itself to reach the next floor using a radio connection to the building's elevator control system.

Artificial Sleuth

CyberGuard 3's scanners and environmental sensors are mounted at a height of 1.8 meters, so it can detect less dense polluting gases which rise. An integrated video system enables visual information obtained from the environment to be recorded and transmitted to a control station.

Infrared and microwave sensors with a range of 360° detect intruders. At room temperature, these infrared sensors are able to perceive human body warmth at a range of 20 to 50 meters. The microwave sensors function at a frequency of 25 gigahertz and are able to spot the tiniest of movements made by humans and equipment 20 meters away.

A flame-recognition sensor can register the flame of a cigarette lighter up to 5 meters away and therefore, together with smoke and temperature sensors, the mobile robot is a very suitable fire detector. Furthermore, it is possible to equip the vehicle with a variety of gas sensors to monitor concentrations of carbon

Monitoring sequence recorded by Cyberguard 3, Cybermotion, USA

monoxide, acetone fumes, methane, and other gases in the air. The first commercially implemented CyberGuards are performing their duties at the Los Angeles Museum of Art. Other devices have been delivered to pharmaceutical companies and to the U.S. Ministry of Defense.

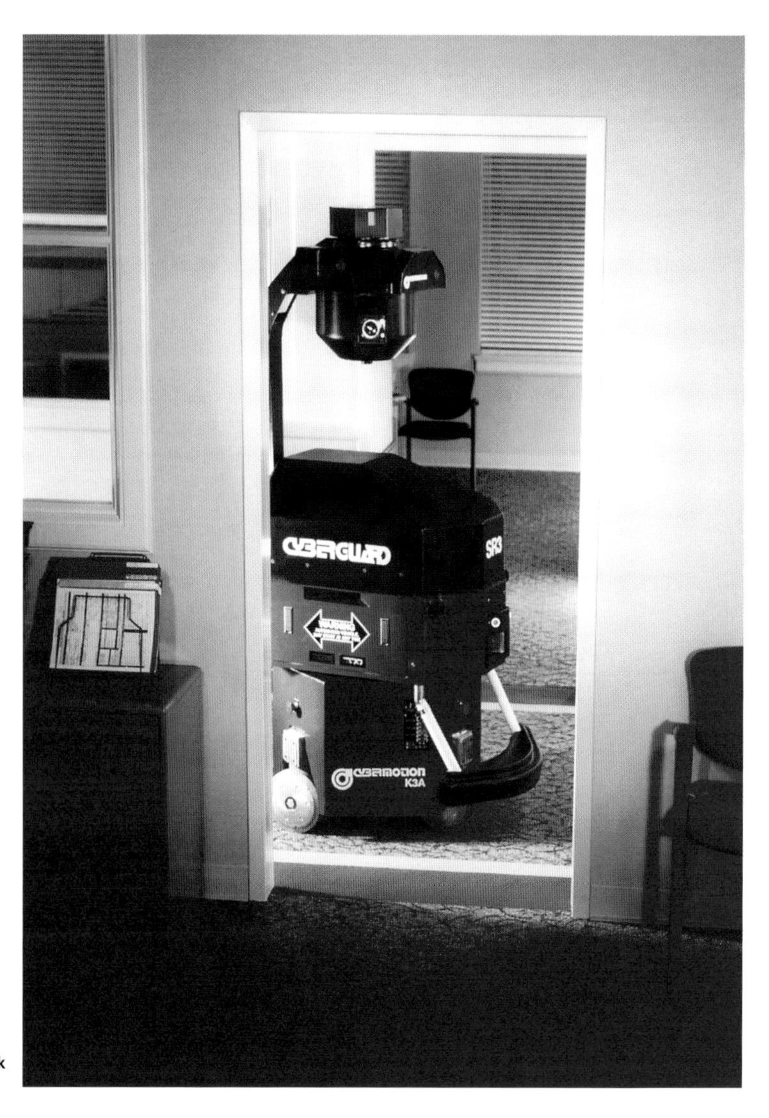

Cyberguard 3 at work

Observant Dragonfly

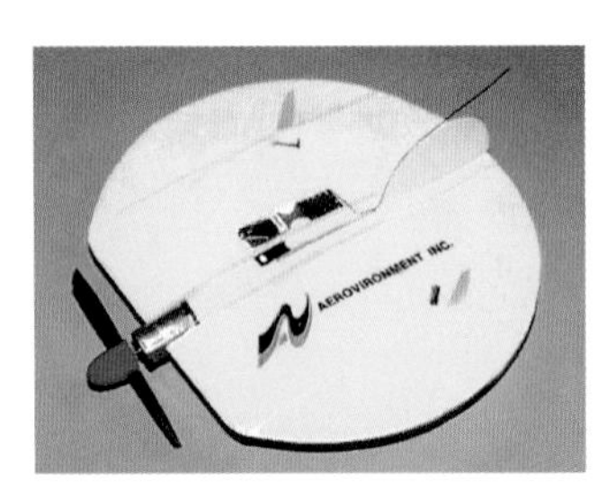

Prototypes of the Micro Air Vehicles (MAV) made by Aerovironment, USA

Micro Air Vehicles (MAV) seem to have come from a Bond film: The mini-aircraft, developed by Aerovironment of Simi Valley, California, represent the limits of current technical feasibility in the area of miniaturized air reconnaissance.

Development of these new craft began in April 1996 with the conception of a miniaturized remote control. In October 1996, a flying object 45 centimeters long and weighing only 39 grams had its first flight, which lasted for a full two minutes.

On the basis of this success, a six-month project was initiated in October 1996; the aim of the project was to clarify the suitability of micro-aircraft for military reconnaissance work. To this end, the flying device's various subsystems were tested to determine their usability.

Electromotors and internal combustion engines were also available as power alternatives.
The most varied of wing dimensions were also tested, as were propeller variations and remote-control systems. With the aid of a morphological procedure, all the potential solutions to three relevant practical military tasks were measured and evaluated. It was decided at the end of the program that the three best solutions should be produced on a scale of 1:1.

Black Widow with GPS

Two of the aircraft, now reduced to 15 centimeters in size, were powered by a propeller-electromotor combination and had a remote-control unit weighing just two grams. The wing design of one craft was conventional, but the other flying object looked like a disk. This disk-shaped micro-aircraft was given the name "Black Widow." The third prototype, which was not controllable, was powered by an internal combustion engine.

Aerovironment's Black Widow can stay airborne for over 16 minutes; that's not in the laboratory, but outside! The remote-controlled battery-powered mini-plane can reach speeds of nearly 70 kilometers per hour.

The next generation of Black Widows should redefine technical feasibility for micro air vehicles. Their design engineer, Matt Keennon, also wants to integrate a GPS (Global Positioning Satellite) navigation system, sensors for stabilizing flight automatically, a magnetic compass and an altimeter, and a speedometer. The first plans have already been drawn up.

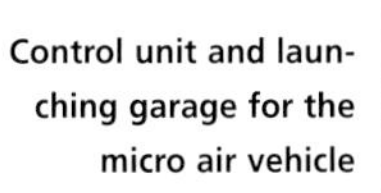

Control unit and launching garage for the micro air vehicle

The micro-motors manufactured by the Swiss company RMB are just what is required. The smallest industrially manufactured electro-motor weighs only 0.3 grams!

Another notable engineering feat is the integrable video camera. It is the size of a sugar cube and, together with a transmission switching circuit the size of a postage stamp, forms a video camera with a total weight of 6 grams. This system has already been successfully tested on the 45-centimeter aircraft.

It remains to be seen whether such developments will be allowed to prove their worth in civilian use.

The high-tech components which make up the flying device

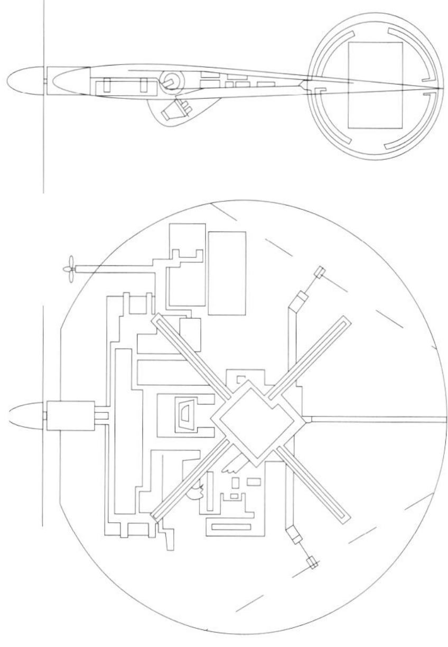

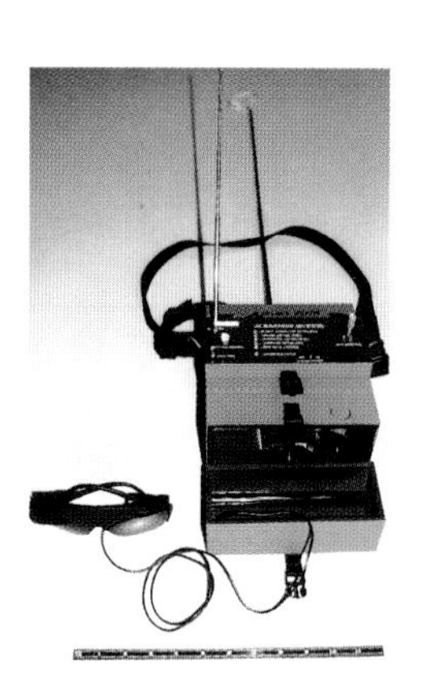

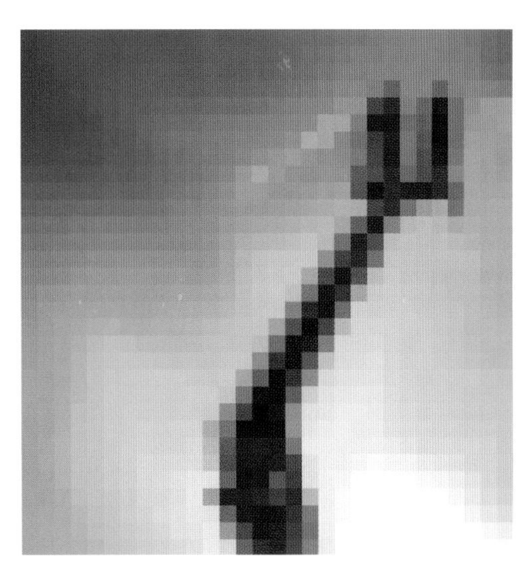
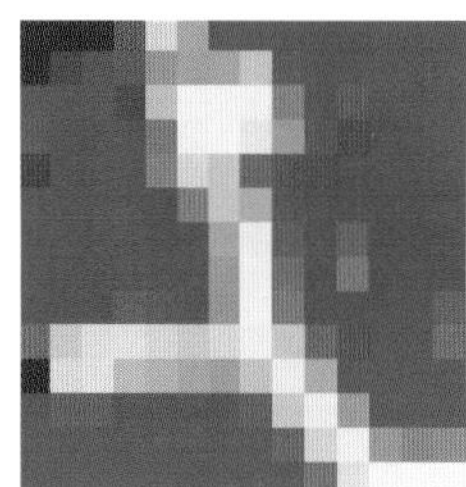
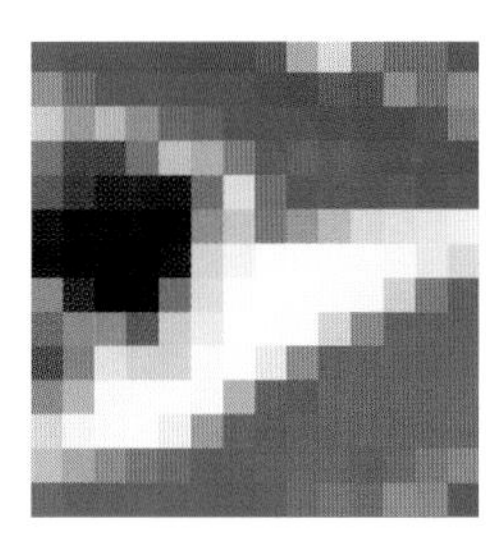
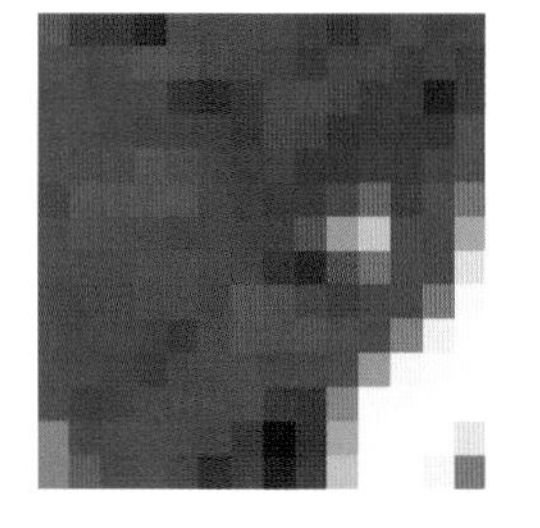

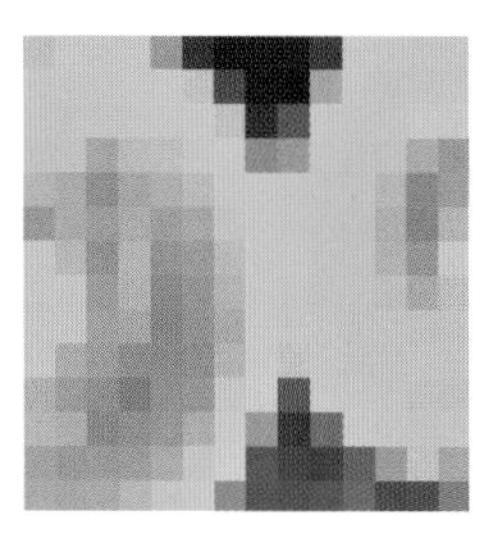
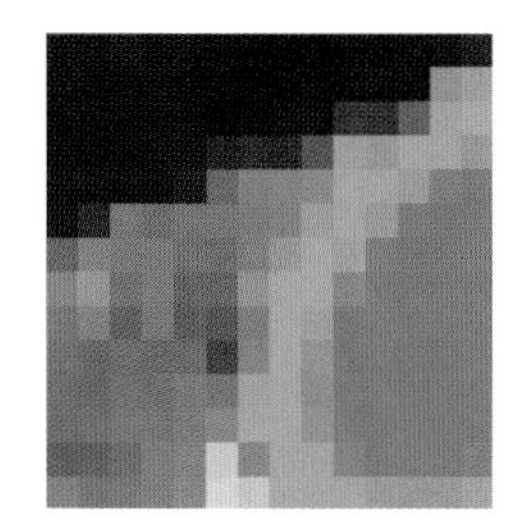

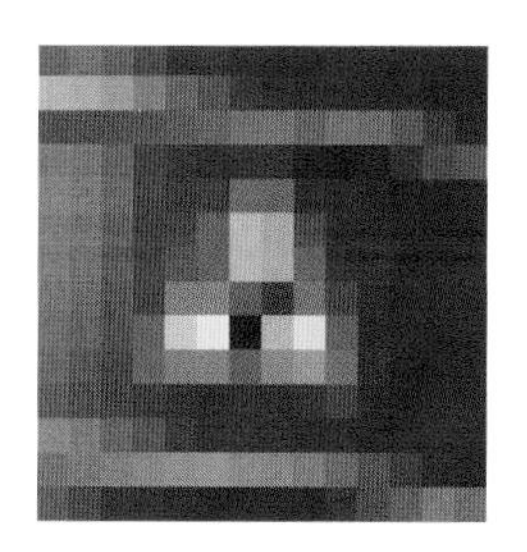

Firefighting

Fleeing from the Flames

Ever since man has known about fire, he has also known of its destructive force. Even today, anyone extinguishing a fire and rescuing others from the flames is still putting his or her own life in danger.

Firefighting bomb, 1715

The invention of the firefighting bomb in 1715 was the first attempt to extinguish a fire from a safe distance. This device, consisting of a water-filled barrel loaded with two pounds of powder and a fuse, was rolled into a burning building and detonated.

Only in this century has it become possible to fight fires from a safe distance using remote-controlled machines. The first systems were developed for fighting aircraft fires. Rescuing people from burning planes is a particularly dangerous task because fuel that has leaked and ignited produces a powerful wall of fire, separating the rescuers from the victims!

Using Foam to Create a Breach in the Wall of Fire

In 1968, Professor Ernst Achilles and Dr. Oskar Herterich designed a small, tracked remote-controlled carrier for extinguishing agents which was transported by an airport fire truck to the site to fight fires occurring at airports. Once on site, it left the transport vehicle, traveled independently to the burning aircraft, and used foam to make a breach in the wall of fire.

The extinguishing hose attached to the fire-fighting robot was 120 meters long, and rolled up on a drum in the transport vehicle so that extinguishing agents could be supplied even as the hose was being unwound. The winch on the supply vehicle's hose drum wound the hose back immediately if the firefighting tank reversed.

Based on the Goliath tank, a model from World War II, the vehicle weighed 850 kilograms and was 1.2 meters wide, and, including the extinguishing monitor, 2.8 meters long. The remote-controlled vehicle had its own energy source supplied by a battery. It was equipped with several foam monitors with which a breach could be made in the fire wall in the easiest way.

foam monitor
A foam monitor is a jet hose which produces foam made from a mixture of water, foaming agents, and air, which smothers fires under a foam blanket.

Vision of fighting aircraft fires with the aid of a firefighting robot, 1968

Florian Flies Like a Rocket

The cabin of an airplane can only withstand a fire caused by burning fuel for around 130 seconds. Even 1,000-horsepower fire trucks traveling at 110 kilometers per hour cannot get to the burning aircraft that fast. For this reason, Professor Ernst Achilles, an internationally renowned expert in firefighting and former director of the fire brigade in Frankfurt, campaigned in 1971 for the implementation of firefighting rockets for fighting aircraft fires.

Fighting an aircraft fire with extinguishing foam

On the basis of existing rocket technology, Achilles, together with Honeywell in Offenbach, Germany, developed Florian, a "firefighting rocket system for use in the vicinity of airports". The rocket was able to drop its 50 kilogram load of extinguishing agents in flight at a defined point in its trajectory.

The location of the burning plane was determined either from the tower or another command point; within seconds, the trajectory and drop time of the extinguishing agents was calculated and the remote-controlled rockets were then launched. The containers of powder were released above the target and opened exactly over the center of the fire. They contained the fire until the airport fire brigade was able to continue firefighting with its vehicles and rescue the passengers.

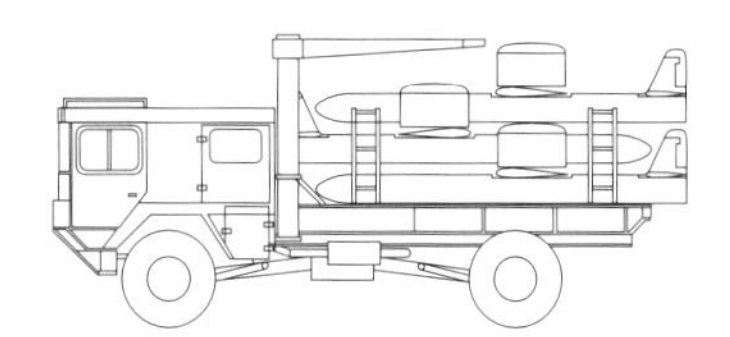

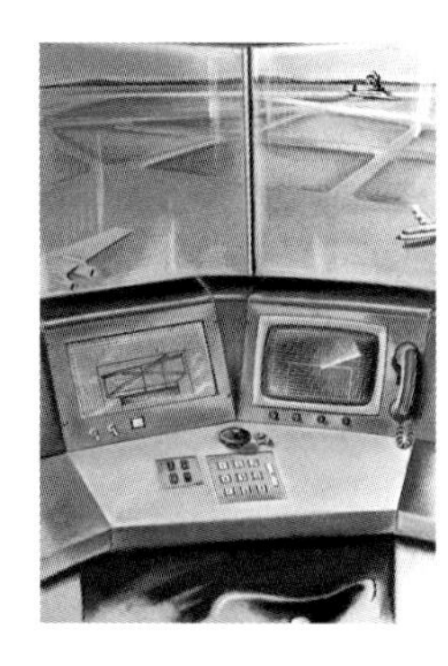

A vision of extinguishing a fire using firefighting rockets

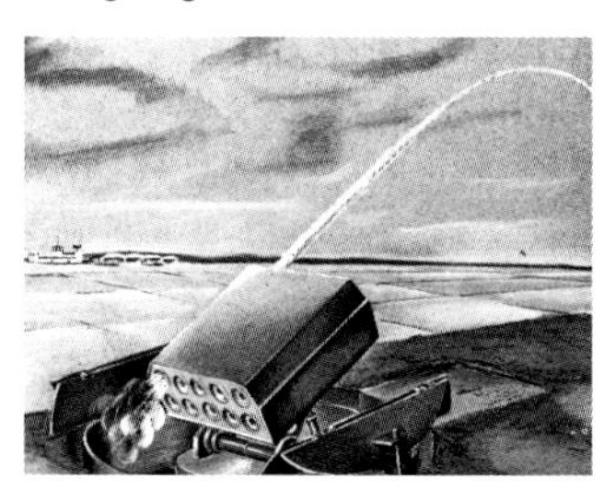

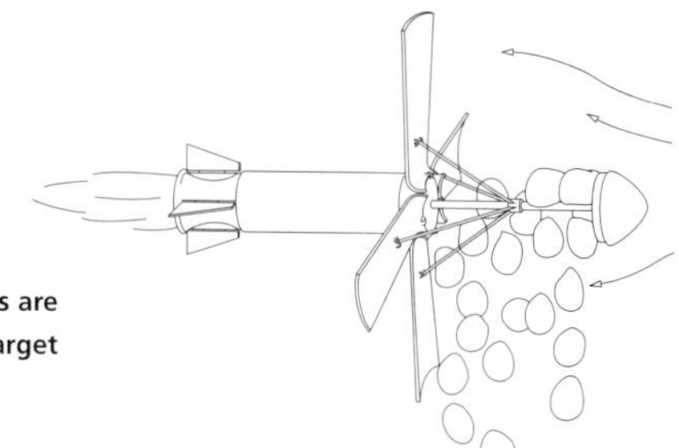

The firefighting packages are released over the target

Modeled on the German V 1

Florian's payload of 50 kilograms was too small, however. Therefore, in 1972, Achilles and the ERNO Raumfahrttechnik company in Bremen, together with manufacturers of firefighting agents, designed remote-controlled flying devices able to carry 500 kilograms of extinguishing agents.

All passenger aircraft would be equipped with emergency-frequency transmitters so that the signals emitted from them could guide the flying firefighting missiles automatically to the crashed plane. The lengthy search and waiting times would no longer apply. A camera in the missile could radio an image of the site of the accident to the control point.

A dozen of these unmanned rocket missiles could be fired from rocket launchers in quick succession and in all weather conditions. The firefighting rockets would fly at 100 meters per second and be guided by autopilot to the burning object, where they hurl their 500 kilograms of extinguishing agents from a height of 30 meters into the flames. They would have an operational radius of 8 kilometers. From the control point, it would be decided where the firefighting rocket should land using its parachute. The appearance of the 4.6-meter-long missile was reminiscent of the first German rocket, the V 1; fully loaded, it weighed 750 kilograms. To date, however, the concept has not been realized.

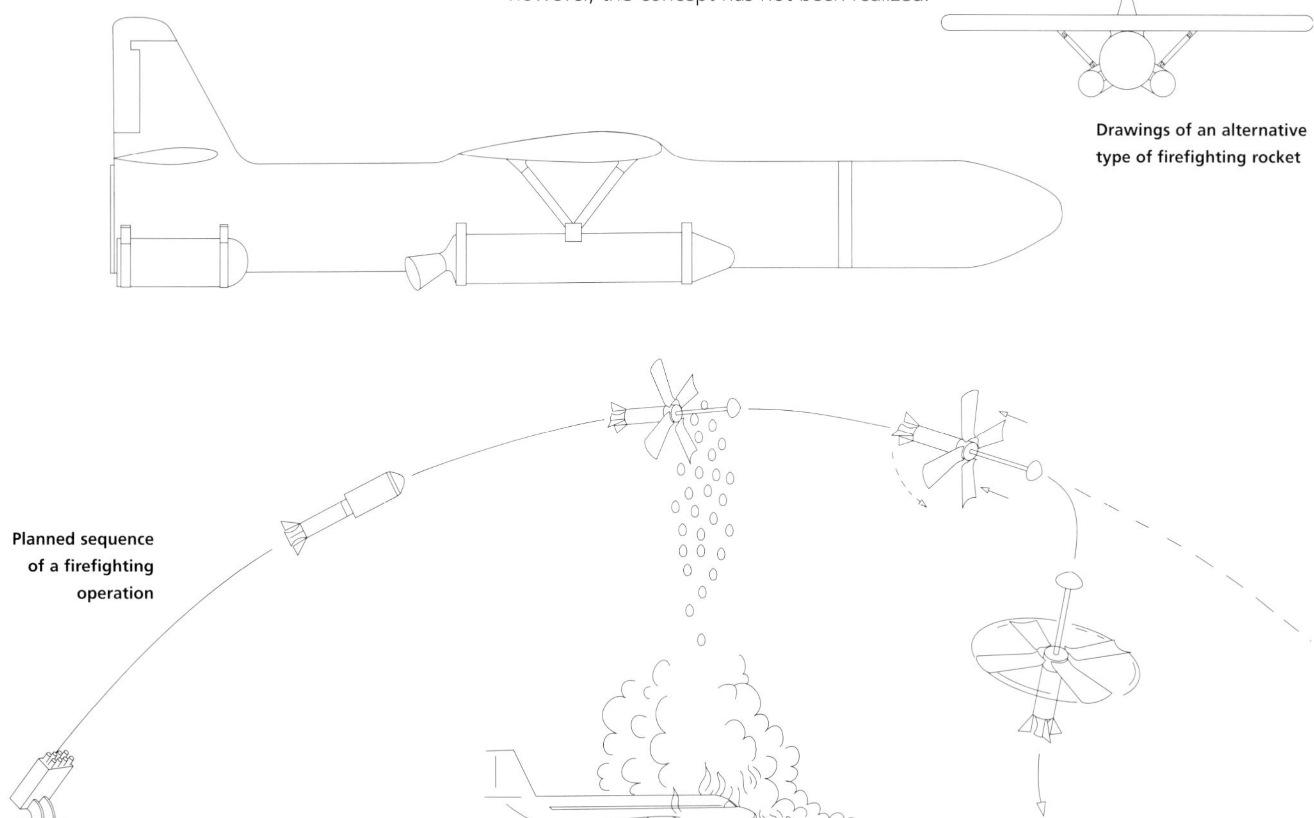

Drawings of an alternative type of firefighting rocket

Planned sequence of a firefighting operation

JetFighter, Japan

extinguishing monitor
An extinguishing monitor is a swiveling jet hose used by the fire brigade that is mounted on a base; the aim of the extinguishing agent emitted from the hose can either be adjusted manually by a fireman or by remote-control.

International Developments

In the USA in 1968, General Dynamics developed a prototype, the FireCat, a remote-controlled vehicle for fighting hazards; in 1973, General Dynamics applied for a patent for the design. The mini-firetruck, controlled by radio signals, was based on a tracked vehicle 1.20 meters in length. Even the aim of the jet hose could be remote-controlled. It was intended that the vehicle would pull the water supply hose behind it; however, it did not have quite the necessary strength for this, so developments were not continued.

In 1972, the Parisian professional fire brigade put a high-performance extinguishing monitor called Sparfeu into service. Sparfeu was mounted on a motorized, remote-controlled trailer chassis. In the same year, a firefighting robot was introduced which was able to enter a burning house with a water cannon and a video camera. The vehicle, weighing around 6,000 kilograms, could reach speeds of 15 kilometers per hour, and its toothlike wheels could even master stairs.

Heat-Resistant for Reconnaissance and Rescue

A research department of the Tokyo Fire Brigade has developed several robots: Rainbow 5, FireSearch, WaterSearch, and RoboCue. The purpose of Rainbow 5 is to fight large-scale fires which develop extremely high temperatures and carry explosion risks. This robot is particularly suited to fighting oil fires, such as those occurring in aircraft catastrophes, accidents involving tank trucks, and accidents in petrochemical installations.

Rainbow 5 can be controlled either with or without cables. Several TV cameras, one of which is equipped with 3-D technology, permit the fire to be monitored. The extinguishing agents are emitted at a maximum rate of either 5,000 liters per minute for water or 3,000 liters per minute for foam. Rainbow 5, however, needs a lot of room at the site where it is used because the device is relatively large, being 4 meters long and 2 meters wide. Because the hoses supplying the robot are pressurized, the robot's ability to maneuver is also limited.

The small, tracked robot, FireSearch, is used exclusively for fire reconnaissance. It has a variable traveling height and is equipped with searchlights, swiveling TV cameras, temperature sensors, and a gripping arm. Underwater, WaterSearch finds and rescues people in distress at sea and assists divers in salvaging tasks. Another climbing robot is used for fighting fires in tall buildings. It can even cut panes of glass.

RoboCue's domain is rescuing people from large buildings. It gets people out of burning factories, department stores, warehouses, subway tunnels, and areas where there is a risk of explosion, such as aircraft catastrophes and chemical plants. The robot is equipped with searchlights and several TV cameras for this.

Rainbow 5, Japan

RoboCue, Japan

FireSearch, Japan

Where visibility is impaired due to smoke, both an infrared search system and contact sensors assist the video cameras in finding people. Loudspeakers and microphones allow controllers to communicate with the people needing assistance; an integrated air canister even enables these people to breathe fresh air.

The firefighting robot made by the English company Ai Security is also a tracked robot platform with an extinguishing monitor mounted on it. Water and energy are supplied by two separate hoses pulled by the vehicle. The platform can be utilized as a carrier system for other purposes, such as use as a manipulator.

Fires Endanger Human Firefighters

Firefighting robots are needed if a situation would be too dangerous for humans. Such situations would include aircraft fires, buildings in danger of collapsing, incidents involving the transportation of hazardous goods, and containers that are leaking or threatening to burst.

To date, the chassis of such firefighting robots has often been a small-sized tank used by the armed forces or former manipulator vehicles designed for use in nuclear installation catastrophes. They are utilized for detecting gases and for simple handling tasks and do not usually possess a heat shield.

Robots used until now have an extinguishing monitor integrated into them. Because of this, additional robots are needed to fight large fires as the extinguishing capacity of one monitor is inadequate. The high purchase price of a robotic platform (from about $55,000 upwards) makes it impossible for a fire brigade to own the several robots required to fight large-scale fires.

WaterSearch, Japan

MF-2, telerob, Germany

Future Concepts Need to be More Reasonably Priced

Based on experiences to date, the company Iveco Magirus Brandschutztechnik GmbH in Ulm, Germany, together with the German company telerob in Kiel, has developed a new concept for a firefighting robotic system that can deliver several trailers, each containing a reel and an extinguishing monitor.

The system consists of a tracked, teleoperated MF-4 robot manufactured by the company telerob. It has a long-range extinguishing reel with a semi-rigid rubber hose and a remote-controlled extinguishing monitor, which can be stowed outside the rear compartment of a standard German firetruck LF 16/12 instead of a DIN 14826-2-type one-man reel.

The MF-4 robotic system is used all over the world by armies, police forces, and security guards for handling suspected explosive containers or objects safely. The MF-4 vehicle has a heat shield that can even withstand a flashover. The interior is cooled with CO_2. The tracks are also heat-resistant.

The hose reel transports a semi-rigid, two-chambered rubber hose wound on it, and a control and supply cable for the monitor. The aim of the monitor is adjusted by remote-control from the supply vehicle using a joystick.

One Robot for Several Monitors

The mobile MF-4 robot is used only for transport purposes. The actual operational firefighting unit is the long-range extinguishing reel with an electrically powered monitor attached to the pulling vehicle as a trailer.

The long-range extinguishing reel is pulled by the robot to the site where it is needed. In the processes, the hose, which is supplied with wate from the fire engine, is continually rolled out from the reel. The radio-controlled robot positions the extinguishing monitor towards the center of the fire at the site and uncouples the reel.

MF-4 then leaves the area of danger immediatel so that it can position additional reels at other suitable sites. In the case of operationa sites further away, several hose reels can be connected to MF-4 simultaneously. The reel o the monitor can also be brought into position by appropriately protected firefighters withou using the robot.

Should a building collapse or an explosion occu only the comparatively cheap hose reel with its monitor and hose material is lost. The expensive mobile firefighting robot is usually outside the area of danger, and therefore remain undamaged. The MF-4 robot is also available for other activities, such as carrying out measuring, testing and observation tasks, and can also be used as a manipulator-carrying platforn

Within the scope of a research project, Iveco Magirus, telerob, the BASF Company Fire Brigade and the Fraunhofer Institute, IPA, are in the process of further developing the existing prototype of the firefighting robotic system. The system needs to be more practical and be as eas as possible for firemen to operate. It must also be so reasonably priced that all professiona fire brigades and, eventually, all main fire brigade district control centers will be able tc afford one.

flashover
When a fire occurs in an enclosed space, smoke collects near the ceiling. It still contains large quantities of inflammable material that get increasingly hotter. If sufficient oxygen is in the air, the overheated smoke suddenly ignites. The burning smoke heats any objects in the room which, in turn, also catch fire rapidly.

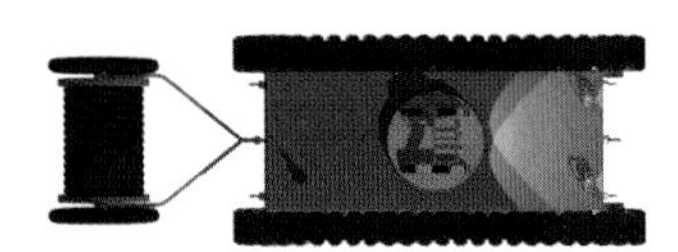

reel
A reel is a two-wheeled frame carrying a drum on which the firefighting hoses are wound. Such a reel enables this type of hose to be transported compactly and rolled out quickly.

MF-3, telerob, Germany

MF-3, telerob, Germany

MF-4, telerob, Germany

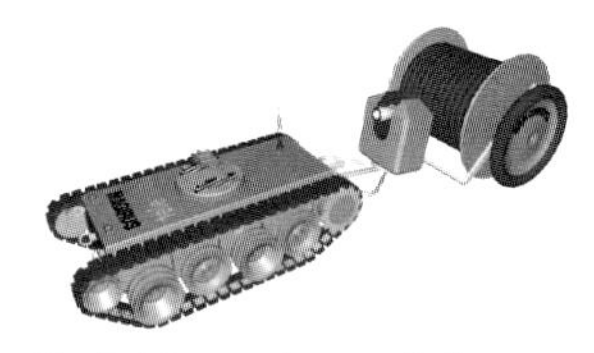

Firefighting system made by telerob and Iveco Magirus Brandschutztechnik GmbH

Platform for the Future

The robot vehicle is equipped with various cameras (visual and infrared), sensors, and control units, and is also able to memorize cartographic information. The extinguishing reel is equipped with a device to detect the center of the fire and automatically adjust the aim of the extinguishing monitor, even if smoke impairs vision. The robotic system can reach the site where it is needed without guidance and can recognize obstacles along the way, even in dense smoke. The intelligent steering unit thus avoids critical situations where the vehicle could overturn.

Because of its navigational capabilities, its intelligent steering unit, its sophisticated sensors, and its optical camera systems, this platform is best equipped to deal with future tasks and further developments in fire protection and can be integrated into autonomous, unmanned firefighting systems.

These vehicles can travel autonomously in industrial areas to dangerous parts of burning buildings or to hazardous-goods containers that are either leaking or threatening to burst. Using teleoperation, extinguishing systems can be positioned automatically and protective measures can be carried out independently on containers of hazardous materials.

These unmanned firefighting vehicles could be stationed at focal points of danger at airports. In the case of an impending emergency landing, the control tower could send the vehicles to the anticipated landing site to initiate the necessary firefighting measures as soon as the aircraft comes to rest.

- Robot positioning an extinguishing reel
- Robot sealing off a leaking hazadous materials transport vehicle with the aid of a manipulator
- Robot waiting in position outside the area of danger

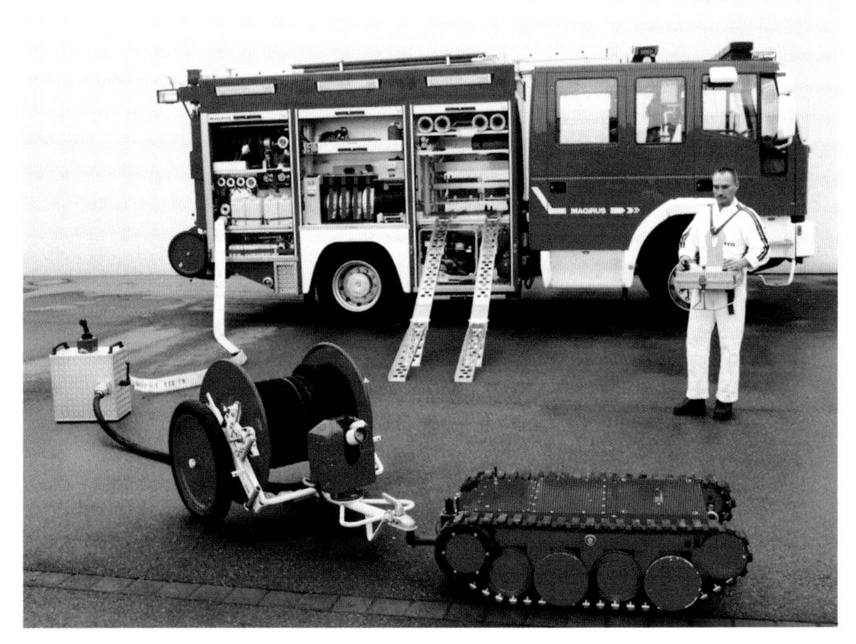

Prototype made by telerob and Iveco Magirus Brandschutztechnik GmbH, Germany

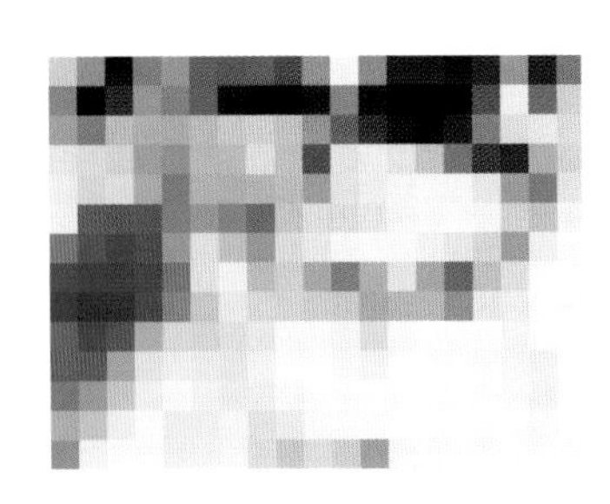

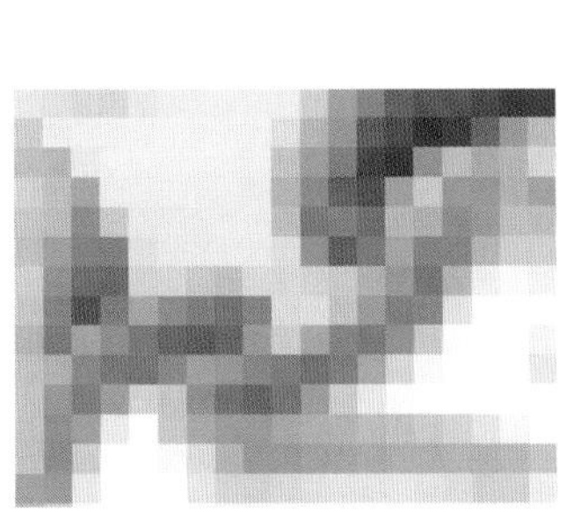

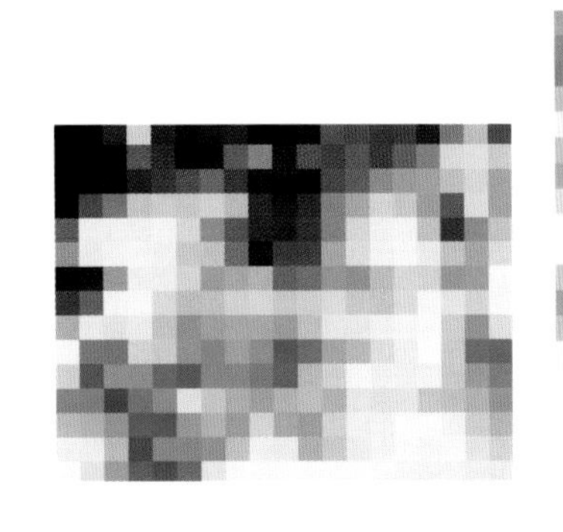

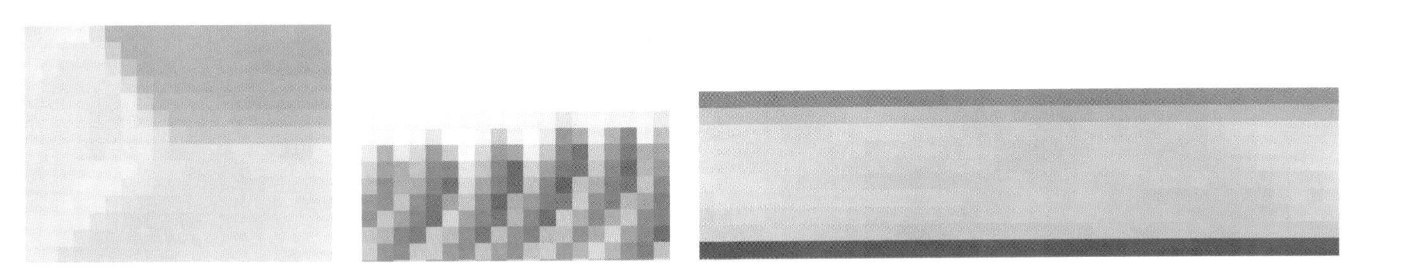

Sorting

Robotic system ULIXES for sorting heterogeneous objects, IMT, Germany

The Goods in the Pan...

Humans live and survive by sorting processes. We sort our impressions of the environment and classify them according to specific priorities. In this way we are able to safely cross a busy street, recognize people, and distinguish between certain materials. From infancy onward, our brains are trained to execute sorting and classifying processes.

This human ability is used today in installations for the manufacture, packaging, and recycling of raw materials and products, and places high demands on automation.

Sorting is Net Value-Added

Sorting processes are absolutely necessary to obtain a certain quality of raw products when extracting raw materials from natural products or mineral sources. In the coal-mining industry in former times, dead rock and coal had to be separated by hand on sorting belts. Since then, the technology for separating materials by physical and chemical properties has been developed sufficiently to allow the process to be carried out on a sorting belt without this monotonous work.

When harvesting natural products, distinctions are made based on size, degree of ripeness, or amount of damage by pests, and products are placed accordingly into categories of quality. In this way, quality norms can be attained, which in turn lead to higher market prices. Sorting, therefore, represents a net value-adding process. When sorting products from a manufacturing process, faulty parts must be recognized and handled appropriately.

Highly Demanding Jobs

Industrial sorting sites are dependent on the speed of the installation's machines because the continuous mass production of goods represents the most economical form of manufacturing. Thus, on bakers' conveyor belts, 60,000 bread rolls must be inspected per hour so that rolls that are blemished or have not achieved the normal size may be removed.

Robots relieve people of monotonous sorting belt jobs. The robot, aided by a task-specific sensor system, is able to recognize the position and orientation of the items to be sorted and picks them up using its specially developed gripper.

A Gentle Grip for Sausages and Chocolates

Robots are capable of sorting all manner of products, from chocolates to geometrically heterogeneous sausages. Manufacturers of robot-assisted sorting systems take skillful advantage of robotic flexibility. When sorting sausages, for example, a robotic system consisting of three co-operating robots is able to sort 10,000 disarranged pieces arriving per hour in such a way that they can be placed in an orderly fashion and be fed into the packaging machine. The speed of a single robot is comparable to manual sorting. Should product quality checks and throughput amounts increase, the sorting problem can only be solved safely and economically by machine. The problem here, however, is that the products are often highly heterogeneous, that is, they cannot be differentiated by just one characteristic.

imt
Wir steuern Ihre Produktion

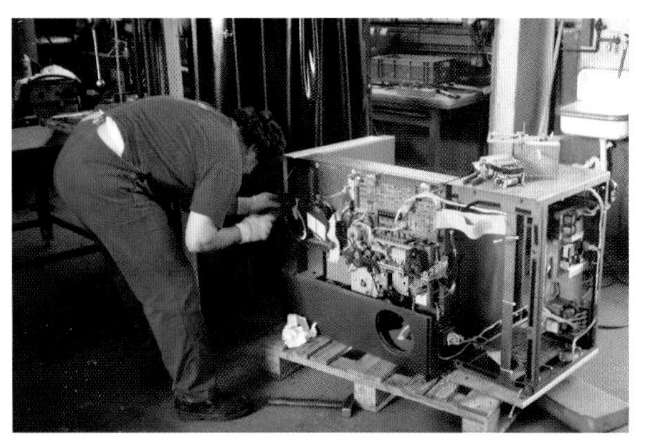

Dismantling electronic scrap for recycling

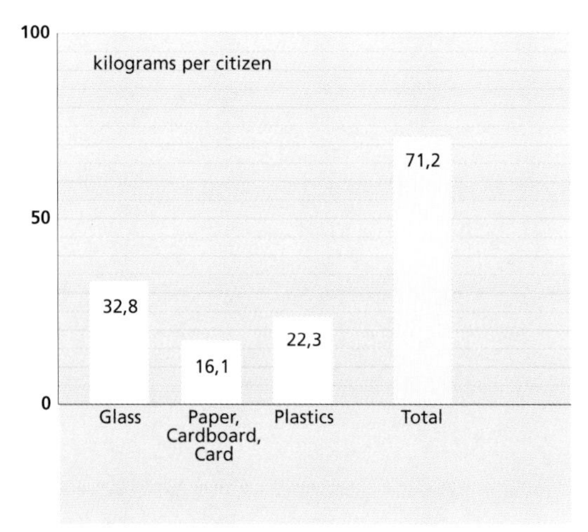

Collection results in Germany

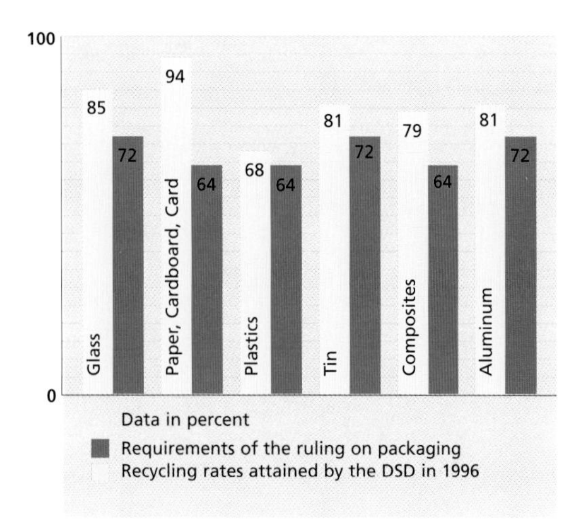

Requirements of the ruling on packaging

Resources from Trash

This problem becomes extreme when sorting recyclable materials, a task that concerns everyone nowadays. In periods of shortage, recycled products can be an important source of raw materials. Today, recycling is achieved mainly through the insight of citizens, enterprises that wish to project an environment-friendly image, or through legislative measures. In 1996, a law concerning product recycling was passed in Germany and thus formed the legal basis for completing the circulation of materials. First, with the ruling on packaging, a product recycling system for lightweight packaging was created so that the flow of amounts in this area could be recorded and recycled. Electronic scrap and used car recycling systems are also being set up at the moment. Sorting is necessary for recovering recyclable materials in all areas of recycling.

To do this, households roughly pre-sort their trash and deliver it to the various recyclable material collecting systems. In this way, e.g. Germany's DSD system, a material collecting system for recyclable packages, amassed around 5.3 million metric tons of recyclable materials in 1996. That equals 84% of sales packaging from households and small companies. The DSD system's logo "Grüner Punkt" has to be printed on almost any package sold in Germany.

DSD
The DSD ("Duales System Deutschland") is an enterprise that is responsible for the recycling of recyclable packages.

Grüner Punkt
The "Grüner Punkt" ("Green Dot") is a logo on the package that shows the customer that DSD's recycling cost for this specific package has been included in the retail price.

Nothing Is More Heterogeneous Than Trash

In the "Gelber Sack" there is a highly varied mixture of materials, including various plastics, drink cartons, and aluminum, which can neither be clearly categorized by amount nor material. Furthermore, because people misplace materials into the wrong containers, materials that cannot be recycled will always be found in the collecting system. Sorting tasks for recycling these materials are therefore particularly difficult to automate: At the moment, sorting still takes place manually using sorting belts.

In addition to the monotony of working in a sorting installation, employees have to work in extremely unsanitary conditions. The transmission of spores and bacteria from the remains of rotting food, and the exposure to nauseous odors and chemical substances add up to form a health risk that has not yet been fully investigated and the long-term effects of which are hard to calculate.

"Gelber Sack"
In Germany recyclable material is collected in any household in a yellow plastic bag called "Gelber Sack". This yellow plastic bag is an integral part of DSD's recycling system.

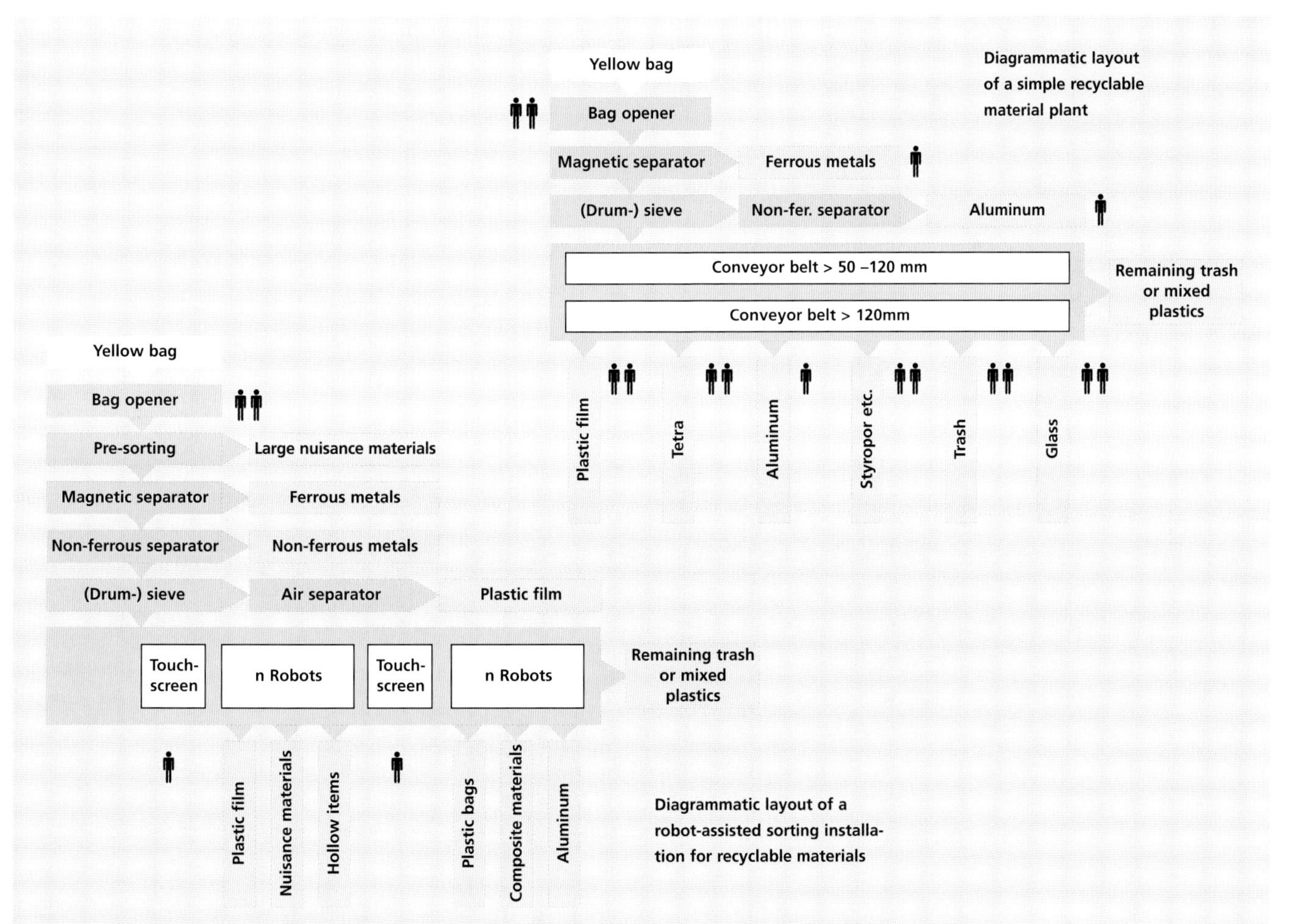

Diagrammatic layout of a simple recyclable material plant

Diagrammatic layout of a robot-assisted sorting installation for recyclable materials

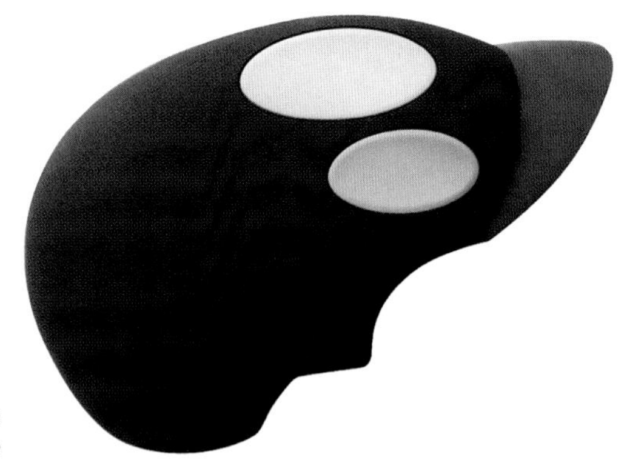

Data collection using a pointer mouse

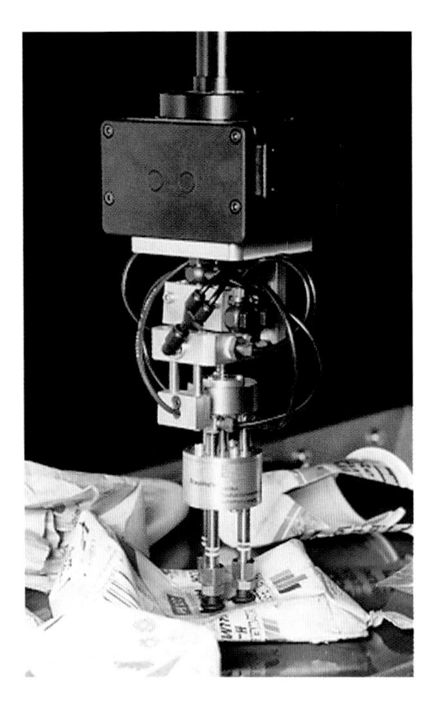

Sorting trash with a special gripping head

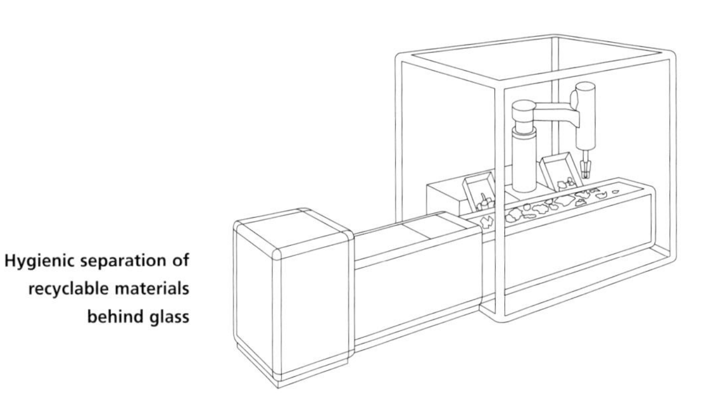

Hygienic separation of recyclable materials behind glass

Man Identifies Materials for the Robot

The composition and state of the recyclable materials have set almost insurmountable tasks for engineers designing automated sorting systems. The items on the conveyor belt first have to be localized, then classified, and finally sorted using an appropriate gripping technique.

In a first step, a partially automated testing cell has been developed by Fraunhofer IPA for sorting recyclable materials. This would clarify fundamental questions concerning sensory analysis, handling techniques, and gripping principles.

First, people must tackle the difficult job of locating and identifying the articles. Using a patented man-machine interface with a transparent touchscreen, a human controls the robot kinematics directly.

Hygiene Behind Glass

The operator interface is mounted as a transparent pane above the sorting belt. By pointing his finger to an object, the operator indicates that this is exactly the object he wishes to have sorted. It is also possible to identify the material by using a specially developed "pointer mouse" and a selection key. In this way, a first step has been taken toward automating a system that shields the operator from the burden on his health caused by airborne microbiological contamination from sorting recyclable materials.

As an example, different grippers equipped with sensors are capable of recognizing aluminum-coated recyclable materials and assigning them to the discharge shaft. Induction sensors react

Data collection via the operator interface

to objects of varying heights and steer the gripping height (z-axis) of the robot accordingly.

Since only standard kinematics can achieve the desired cost-usefulness ratio, the robot itself is constructed like a conventional SCARA industrial robot. In the mid-term, a new design needs to be invented because robots with low levels of accuracy and payload would suffice for this type of application.

Image Processing for Sorting Trash Fully Automatically

The next step toward automated sorting has already been taken by Fraunhofer IPA. A sensor system has been developed that is able to recognize items even if they are lying on top of one another. This is an important step in sorting recyclable materials, as items can only be completely isolated using expensive materials-handling technology. It has not been possible to identify items lying on top of each other using conventional image processing. Solutions to this problem that can also be integrated into already existing installations are now available.

The algorithm developed for this is able to recognize the object to be picked up by its height profile, which has been recorded by a sensor, and can differentiate between one object lying above and another object below. With an appropriately trained neuronal network, it is also possible to identify specific groups of materials. The implementation of robots in environmental and health protection is becoming more and more economical.

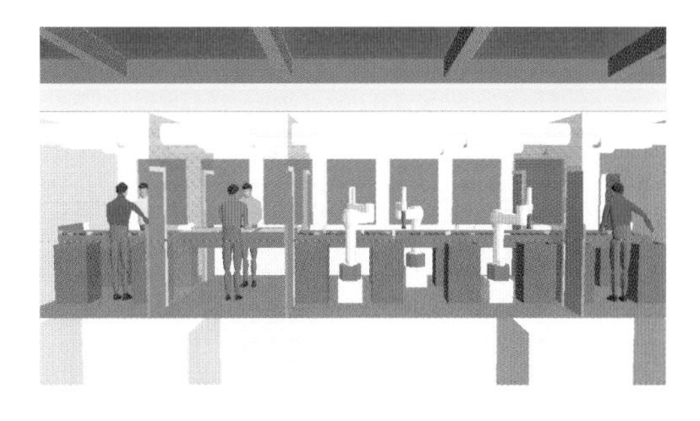

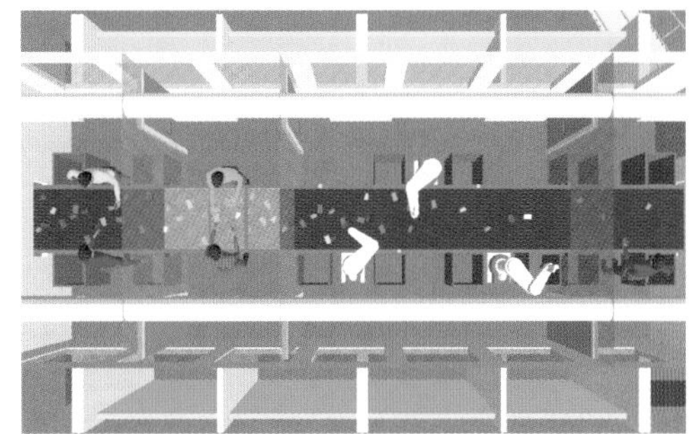

Computer simulation of robot-assisted sorting of recyclable materials

Experimental set-up for sorting trash, Fraunhofer IPA, Stuttgart, Germany

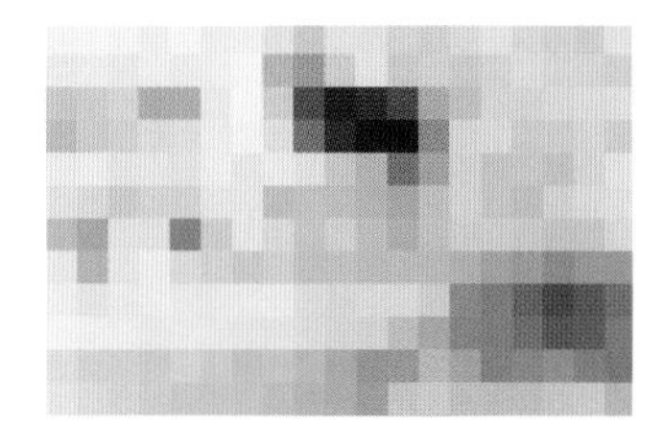

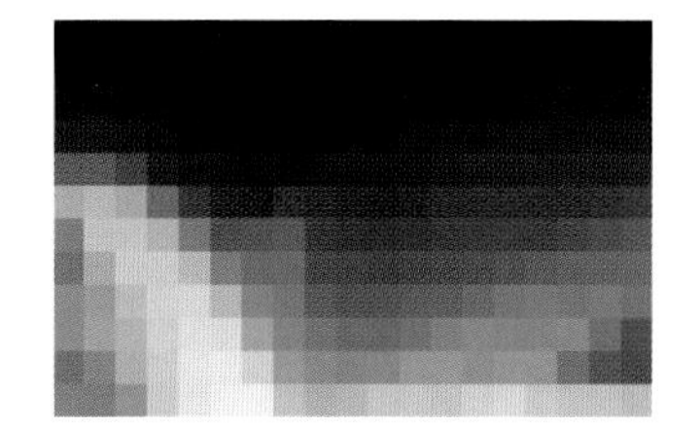 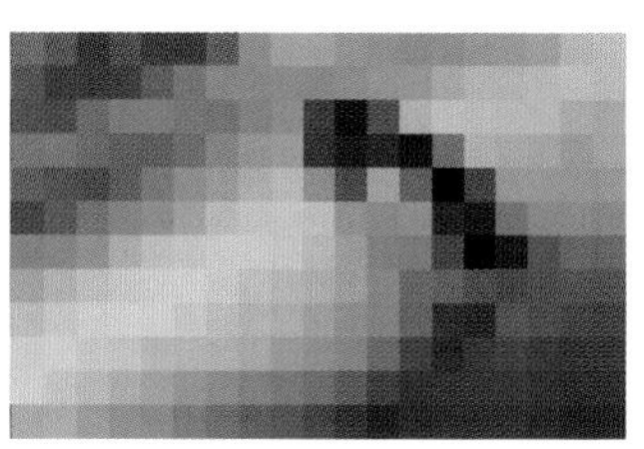 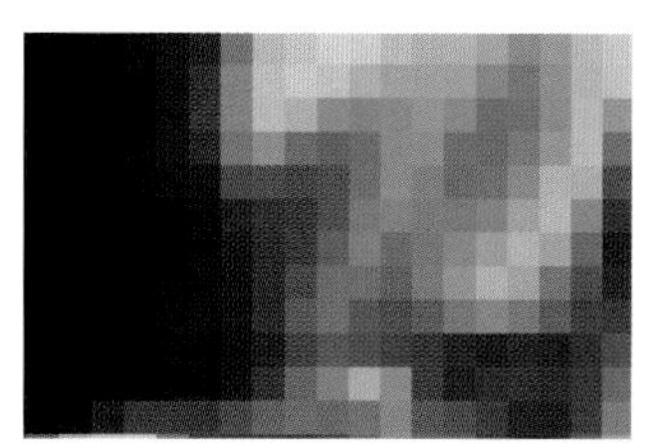

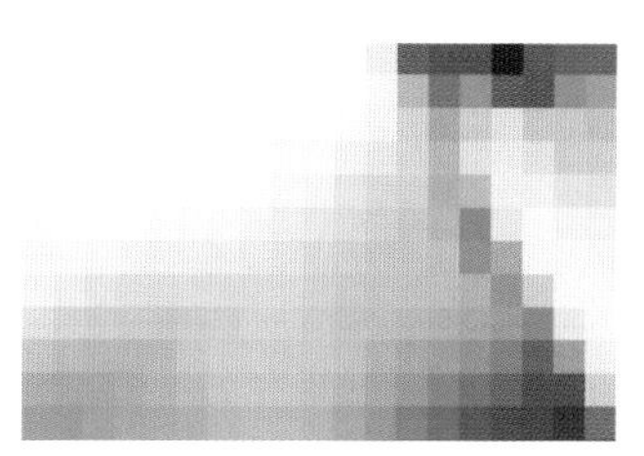

Hotel and Cooking

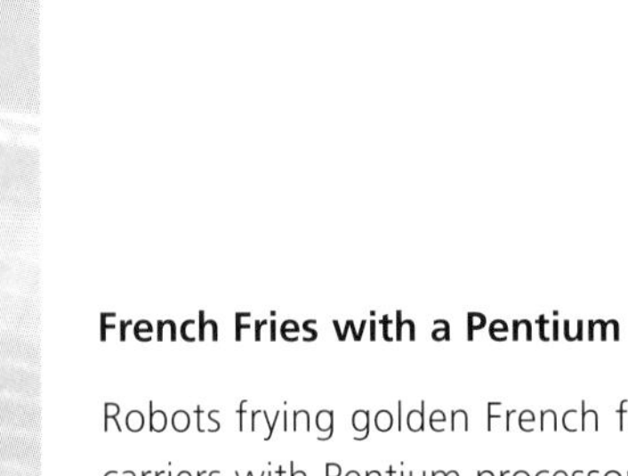

French Fries with a Pentium

Robots frying golden French fries? Luggage carriers with Pentium processors? These are just two examples of service robots in Southern Germany being used for such purposes.

French fries robot, developed by McDonald's, the Frymaster company, and the Gas Research Institute, USA

French Fries of a Consistent Quality

A lot of things can go wrong, even when preparing such simple foods as French fries. Quality is dependent on the temperature and condition of the oil and, also very much on the frying time. To date, the fast-food restaurant chain McDonald's has only been able to assure the quality of the raw products and the oil it uses. The manufacturers of deep-frying pans try to keep the parameters of the frying process constant by measuring and controlling the temperature of the oil.

From time to time, however, a basket of French fries stays in the oil too long. McDonald's, together with Frymaster and the Gas Research Institute, has developed a solution to this problem: the French fries robot. It loads the deep-frying vats automatically.

A Robot Lets Nothing Burn

To do this, the frying basket is first transported by a linear axis to a filling station, where it is always loaded with the same amount of frozen French fries. Next, depending on occupancy, the robot chooses between a maximum of four deep-frying vats into which it immerses its fries. After a specific length of time, as laid down in a quality guideline, it agitates the fries in the basket so that they brown uniformly and do not stick together.

After precisely the stipulated cooking time, the robot transports the fries to a warming container and empties them into it. Here, the French fries are salted by personnel and then packed.

The entire sequence can be either started by an operator or can take place at set intervals. In all, employees are relieved of working at the deep-frying vats and can devote themselves to other tasks or to looking after customers. If a sub-function of the robot breaks down during the hard day-to-day work, the deep-frying vats can also be filled manually. The robot has already been put to use several times in the USA. There is only one experimental application running in Germany at the moment.

French fries robot busy frying McDonald's fries

Mortimer Serves Breakfast

The only service robot in the world working in the hotel industry comes from Karlsruhe, Germany, and answers to the name of Mortimer. At Karlsruhe University's Institut für Prozeß-rechentechnik und Robotik (IPR), under the direction of the physicist René Graf, a robotic butler has been developed which can carry suitcases, serve breakfast, assist with room service, or hand out the mail. Siegfried Weber, a hotelier from Karlsruhe who tested the prototype in 1998, has supported this development from the beginning.

"Robotic butler" Mortimer
IPR Karlsruhe University, Germany
(Photo: Arnulf Hettrich, Stuttgart)

The research project, Korinna (components for robots in innovative applications), gave the impetus to develop this new generation of freely moving service robot. With Korinna, the aim is to develop modular service robots that can be flexibly adapted to their specific field of application. In this project, mobile robots with the most varied kinematics and sensory equipment are produced.

Use in the hotel industry places particularly high demands on a mobile service system. To operate smoothly in this dynamic environment, the service robot has to navigate along narrow corridors, maneuver around stationary and moving obstacles, and interact with people and equipment. Despite these numerous demands, the device has to be reasonably priced.

The research team from Karlsruhe therefore agreed a specific design for Mortimer, their "robotic butler for the hotel trade." Mortimer has an octagonal, symmetrical base with a diameter of 72 centimeters. Four passive stabilizing wheels are firmly connected to the frame of the vehicle and carry the weight of the entire unit. The powered wheels have spring suspension so that they can compensate for uneven flooring. The motor and batteries have been mounted in such a way that they exert a high pressure in all operating conditions and reduce slippage to a minimum during motion. The transport platform is 45 centimeters high, allowing heavy suitcases to be easily placed onto it. The tallest side of the mobile robot measures 115 centimeters.

Multi-Sensor System Identifies the Environment

Mortimer uses a multi-sensor concept with a two-dimensional laser scanner, an array made up of 16 piezo-ultrasound sensors, and a camera system to identify his environment. In addition, eight tactile sensors on the outer surfaces of the octagonal robot help to feel objects at the last second that he would otherwise collide with.

The camera system for identifying features is mounted on the robot's highest point. Together with the two-dimensional laser scanner in the main body of the robot, it creates a three-dimensional model of the environment. The camera image supplements the laser scanner data, and supplies important information that is especially useful for identifying static obstacles such as the edges of doors or other stationary objects.

The measurement data from the laser scanners is used to correct the robot's position and, in conjunction with the measurement values from the ultrasound sensors, to avoid collisions. Because of its mounting locations, the laser scanner is only able to record objects that are about 20 centimeters above the ground. For this reason, ultrasound sensors have been attached to the robot's circumference which, due to their relatively large measuring cone with an aperture angle of 52°, are able to monitor the mobile vehicle's entire environment.

Orientation Without a CAD Model

Mortimer gets his training by following a person around the hotel and thereby learning details about his surroundings; thus, a CAD model of the working area is no longer needed. Mortimer creates and memorizes his own impressions.

Along with the hotelier from Karlsruhe, other companies have also shown an interest in Mortimer. In the near future, we may meet a robotic bellboy during a visit to a hotel that asks for a drop of oil rather than a tip.

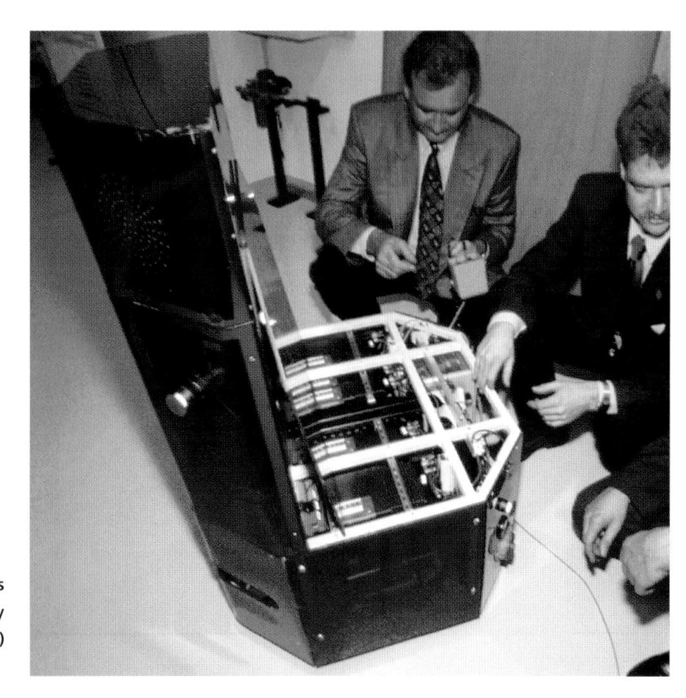

Mortimer's internal workings
IPR Karlsruhe University, Germany
(Photo: Arnulf Hettrich, Stuttgart)

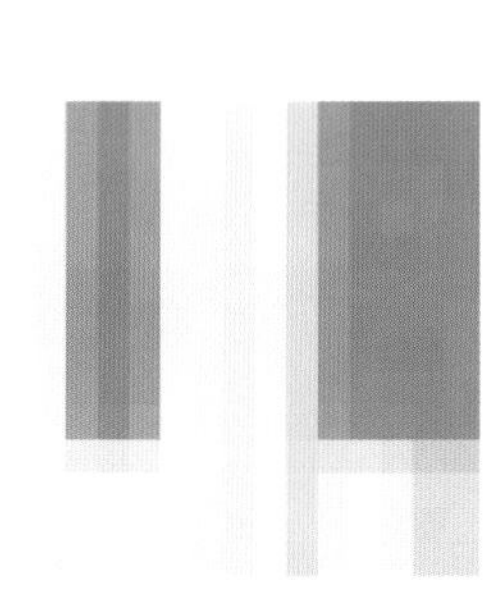

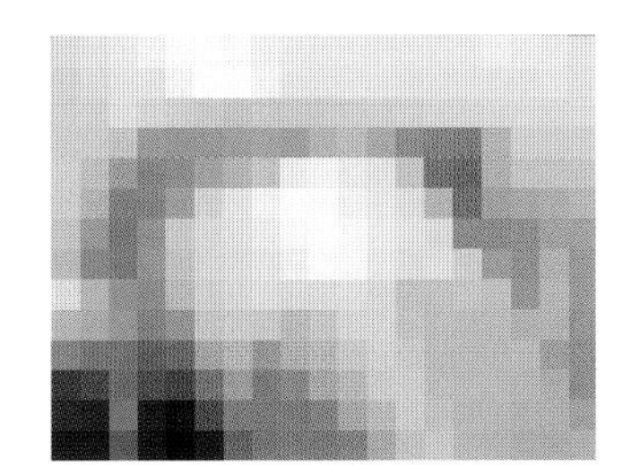
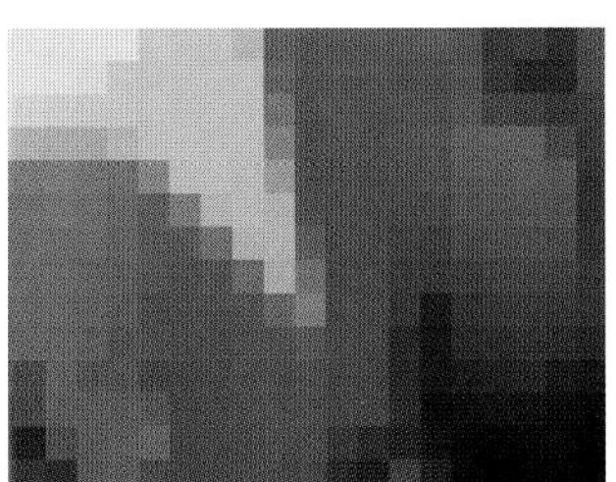
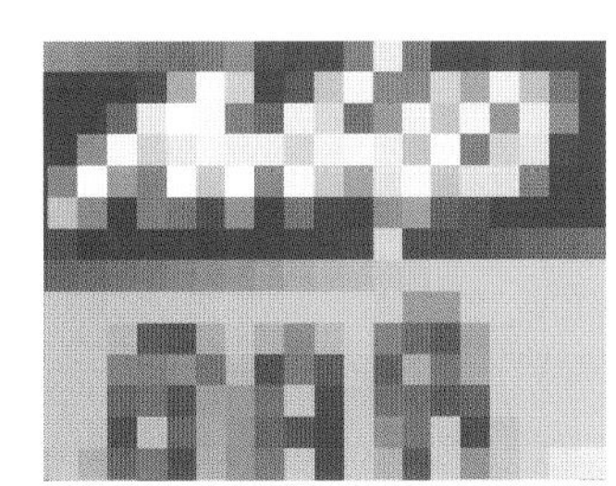

Marketing

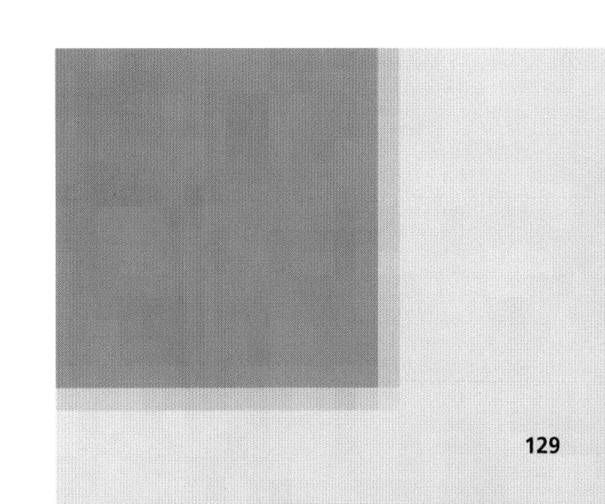

From New Products to Presentations

Consumer products are not the only things that sell better if they cause a big stir when they are presented. Nowadays, the emotional charge surrounding an advertisement or a trade fair is almost more important than the product itself. Advertising budgets running into billions of dollars demonstrate how much companies are willing to spend.

Manufacturers of technological products want to be associated with innovations and the future. For this reason, intelligent systems from the field of automation technology are being used as attractions at presentations.

Robots are Considered Highly Innovative

Eye-catching robotics technology is intended to pique the public's interest at events such as trade fairs. The robots entertain the audience and evoke certain associations that visitors will connect with the product or service.

To be successful at catching people's attention, you have to amaze them. A robot that can tighten screws is boring. If it serves coffee, however, the visitor who walks by at a trade fair is surprised: He or she doesn't expect it. The more unusual the application, the greater the surprise effect. If it is also possible to interact with the system, such as giving it an order or telling it to make certain movements, the visitor will certainly remember it.

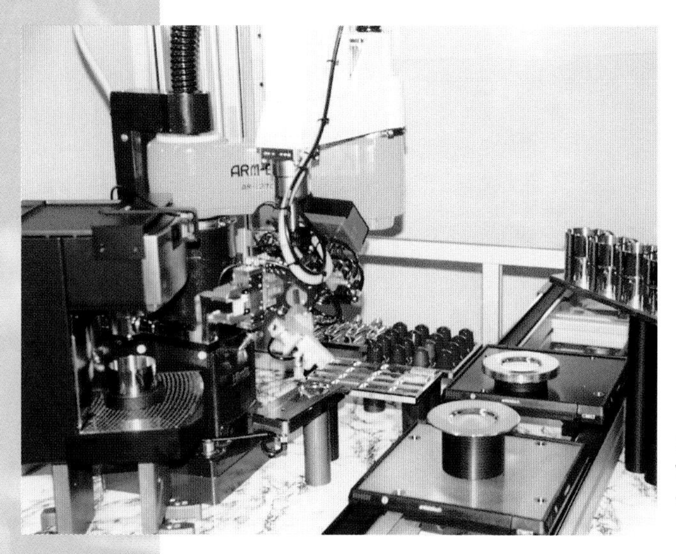

Fully automatic espresso bar, Wolf Produktionssysteme GmbH, Germany

A Special Bartender

The Wolf Produktionssysteme GmbH from Freudenstadt, Germany, enjoys setting up a fully automatic espresso bar to demonstrate complex handling tasks. The robot's pneumatically powered revolving gripper is able to pick up any item necessary, from coffeepot to filter-holder. The robot fills the filter-holder with freshly ground espresso powder and screws it onto the machine. Then it picks up a small espresso cup, places it under the spout, and switches the machine on. While the hot espresso is being brewed, the robot places a saucer and spoon onto a prepared tray. Then it places the full cup onto the saucer, puts a small chocolate with it, and serves the cup to the guest.

For this task, a SCARA robot is used, which has four joints having a workspace of 50 centimeters and a maximum TCP speed of 4.1 meters per second. In this example, the focus of attention was a revolving gripper system for six different grippers. By presenting automation technology in a more unconventional way, the company received a very positive response.

**Fully automatic espresso bar,
Wolf Produktionssysteme GmbH, Germany**

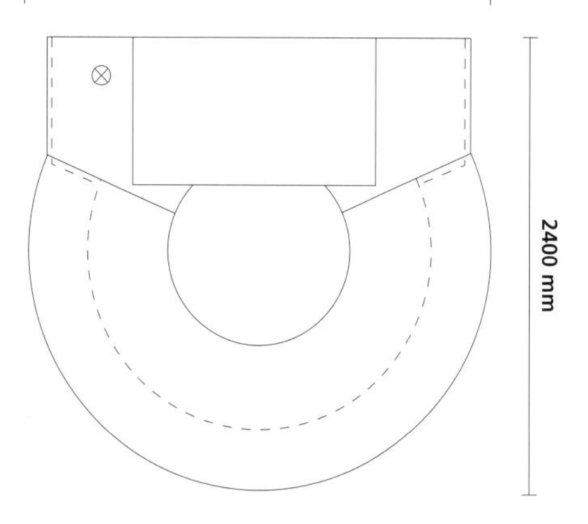

Champagne Waiter with a Delicate Touch

Since 1997, the Erhardt+Abt Company in Geislingen, Baden-Württemberg, Germany, has concentrated on marketing techniques. The company supplies unusual eye-catchers for events such as exhibitions or jubilee celebrations, and designs and creates special applications custom-made for the presentation of particular products or services.

The automatic robot-bar, conceived by Fraunhofer IPA and further developed by two young Swabian entrepreneurs, Stefan Erhardt and Christian Abt, pours drinks and serves them to guests. The robot-bar is made up of a robot, two grippers, and a magazine for bottles and glasses. By pressing a button on the bar, the guest can order a drink. The robot elegantly grips a bottle and two glasses from the magazine and sets them down. It then gently opens the bottle and carefully pours the drink into the glasses.

The robot is particularly fastidious about pouring identical quantities into each of the two glasses and pouring neatly without spilling a single drop. The robot puts the empty bottle down, carefully picks up a glass, and hands it to the guest. Force-sensitive grippers ensure that neither the bottle nor the glasses break.

Industrial production components are used here as well. The bartender is an articulated robot with six joints and two pneumatically powered grippers.

Thanks to its appealing design and the use of high quality materials, these marketing products attract guests at events like magnets.

Champagne robot, Fraunhofer IPA and the Erhardt + Abt Automatisierungstechnik GmbH, Germany

Second generation of champagne robot, Erhardt + Abt Automatisierungstechnik GmbH, Germany

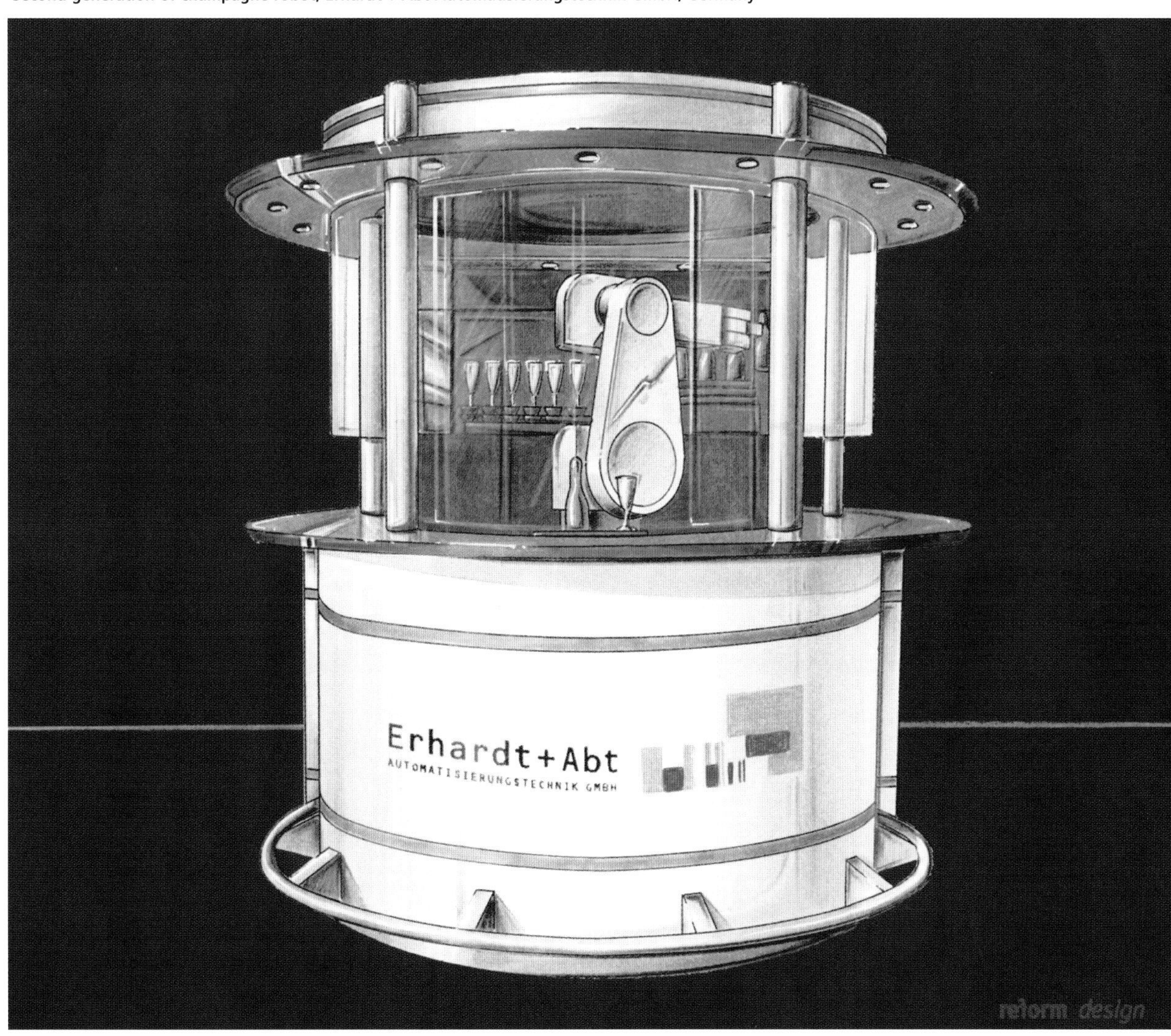

Advertising May Become Mobile

Exterior advertisements are a permanent fixture in the modern world. Neon light advertising and billboards compete with each other for our attention. Affixed to the facades of buildings, these advertisement methods, and others such as clocks and other information systems, are motionless.

The human eye is particularly good at detecting movement. For this reason, advertising specialists are aiming to get people's attention by using moving images or letters on bill-boards or video screens. However, once someone has become accustomed to the position of the information or advertising, as a rule he or she no longer notices it. Purchase prices and operating costs are high for large-sized video screens, such as those seen in stadiums, and are therefore only suitable for advertising or displaying information at mass events.

The effect of habit occurring after a certain period of time with conventional advertising and information displays can be avoided if, for example, the carrier system moves along a facade. Engineers at Fraunhofer IPA, in Stuttgart, Germany, have combined a climbing robot with an advertising or information carrier, and this synthesis will surely gain in popularity over conventional advertising carriers within the next few years. This modular robotic system is called "animated ads".

First Santa, then an Easter Bunny

A conceivable scenario: at Christmas time, Santa climbs around on the facade of a big department store, while at the same time cleaning the window panes and displaying the current temperature on his back. At Easter time, the climber could be given a couple of long ears, and in summer the robot could show itself in its true colors.

Due to the modular type of construction, a whole variety of designs are feasible, all depending on the carrier system and the payload. The carrier system is made up of modules for adhesion, motion, drive, and control. The modules are characterized by their decentralized intelligence. This innovative approach is based on the new fieldbus standard CANopen. The CANopen protocol is a step towards the standardization of CAN communication, and therefore perfectly suitable for this modular type of construction. The frame is made up of high-tensile carbon fiber sandwich tubes. Specially engineered connecting elements provide a high level of flexibility.

The motion module is controlled using a modular control concept based on the programming language C++. This concept can be enlarged upon at any time to include additional elements. An industrial PC serves as onboard computer.

Future work aims at integrating innovative media in the robotic system. Very light displays like E Ink's IMMEDIA will be the basis for new developments. E Ink Corporation, in Cambridge, Massachusetts, has invented an electronic ink display that combines high visual impact, extreme thinness, high curvability, wide viewing angle, and a minimal power draw with an extreme lightweight construction. One meter squared weighs less than one kilogram.

Animated ads – mobile information and advertising carriers, Fraunhofer IPA and reform design, Stuttgart, Germany

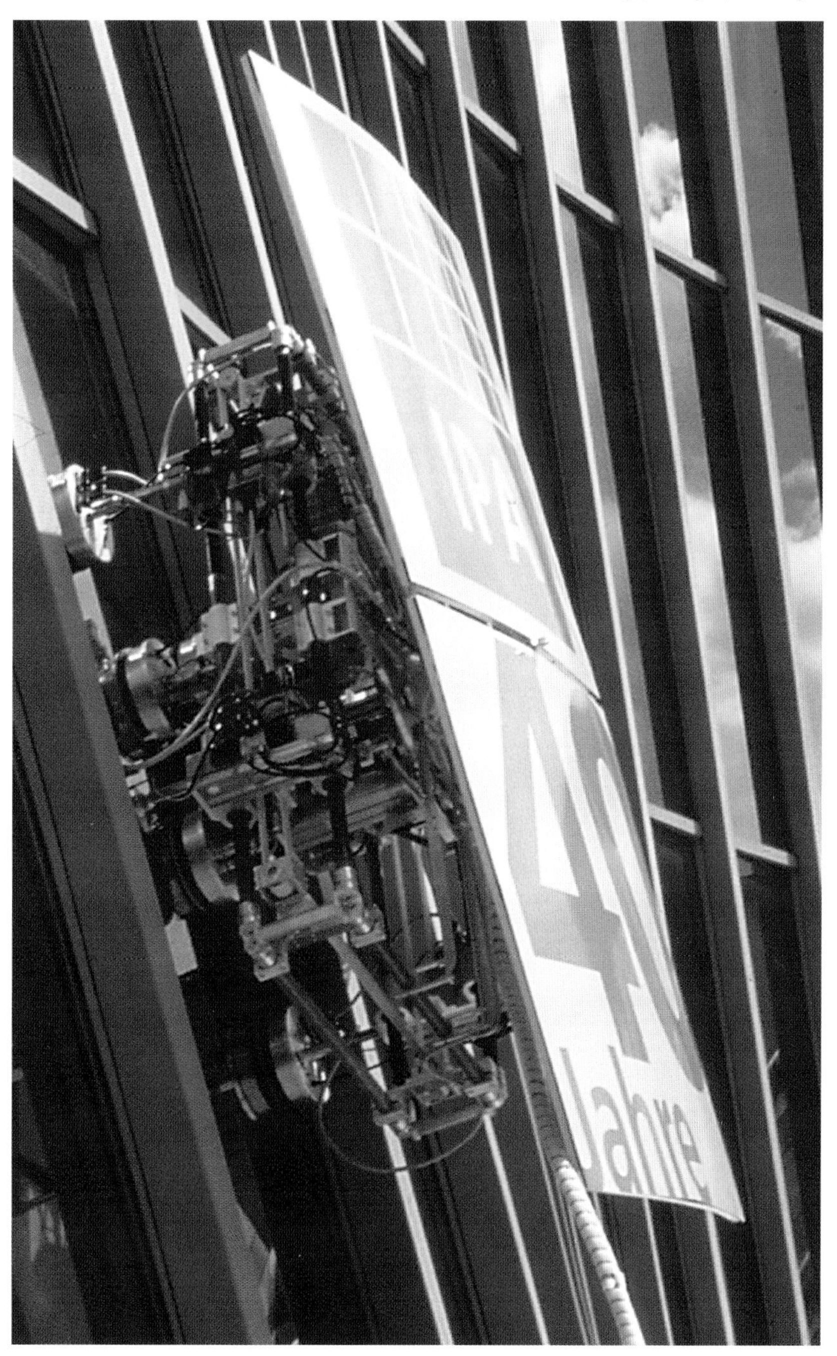

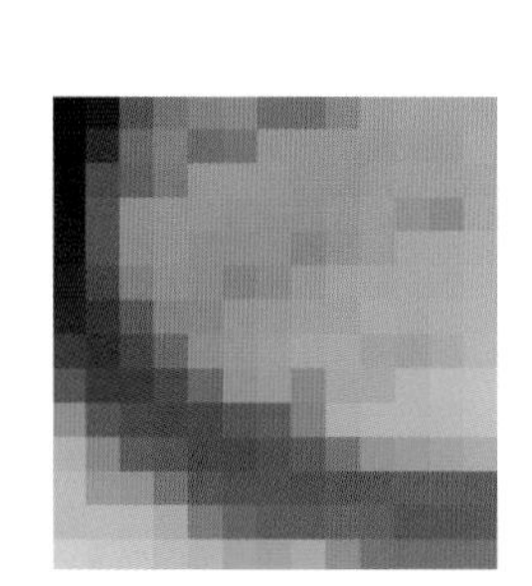
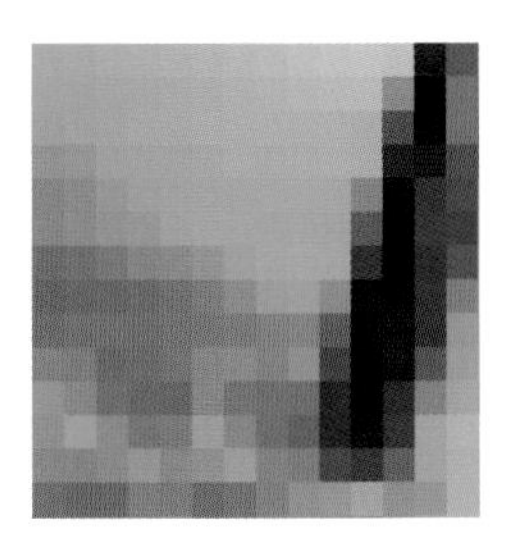

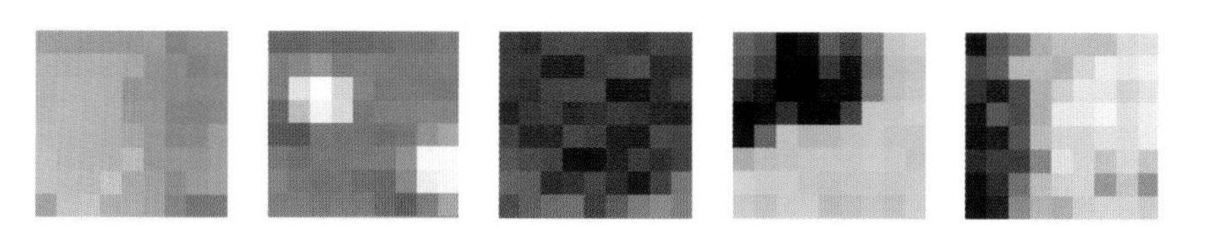

Hobbies and Recreation

Of Sheep, Caddies and Ball-Boys

As a rule, sport and leisure activities are not sensible activities to mechanize. But even here, there are areas where either health or economics is a priority.

Easy transportation of the Solar Mower

Robots Replace Sheep

Inventors from Husqvarna, a subsidiary of the Swedish company Electrolux, came up with the idea of an automatic lawn mower while they were on a seminar in Scotland. On all the lawns around the conference center, sheep were looking after the grass.

"How must a robot be constructed so that it has all the positive characteristics of a sheep, such as always keeping a lawn neatly trimmed and leaving a layer of short-cut grass behind

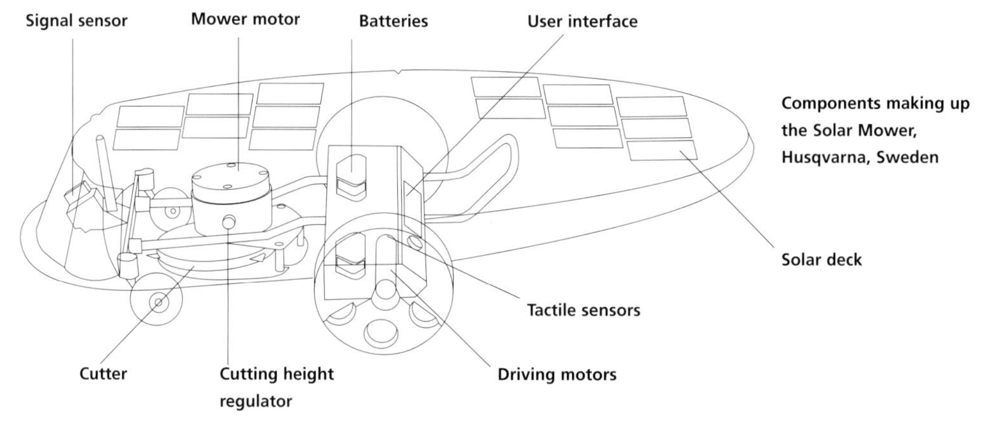

Components making up the Solar Mower, Husqvarna, Sweden

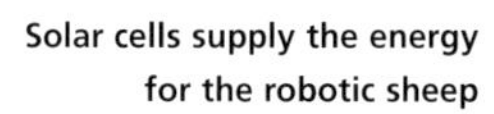

Solar cells supply the energy for the robotic sheep

for fertilization, but do not have the negative characteristics, such as grazing on the neighbor's lawn or even running off?" the participants asked themselves over lunch.

The design engineers from Husqvarna decided on an electrically powered mobile platform that navigates with the aid of induction loops laid into the lawn and tactile sensors built into the vehicle. The small robotic lawn mower, called Solar Mower, obtains its energy from the sun: Solar cells placed on the upper side of the device supply two nickel cadmium batteries with energy.

The Robotic Sheep Lives off of the Sun

The electromotor of the cutting roller has a sensor that measures the torque of the cutting edge. This value is then transmitted to the onboard computer. If the torque drops below a certain limit, the small intelligent helper recognizes that it is moving over an already-cut lawn surface and looks for new areas to mow.

If the device intrudes into a zone that is always in shade, the computer checks how much power is left in the batteries and decides how long it can mow this area before it has to find the sun again.

The induction loop, also powered by solar energy, has to be laid around the entire garden and trees; it also limits the area to be mowed, acting like a virtual fence to the "robotic sheep". It transmits two different signals that are analyzed by a receiver integrated into the robotic lawn mower.

Induction Field Ensures Safety

The first signal is transmitted on a secure frequency, which the lawn mower must be capable of receiving within the entire working area. If this contact is lost, the device stops moving automatically, and the fault causing the signal loss, such as a faulty induction loop, must be repaired before the robotic lawn mower can start working again.

The second signal is transmitted from the induction loop on a different frequency and is responsible for navigation. It signals to the autonomous lawn mower that it is still about 40 centimeters away from the induction loop, and that it is either at the boundary of the land or in front of an obstacle. The robot moves back and turns. Unexpected obstacles, such as garden tables, for example, are only identified when the built-in impact sensors make direct contact. In this way, the Solar Mower can deal with up to 1,200 square meters of lawn.

To prevent the device, which costs around $2,200, from ending up working in somebody else's garden, it is equipped with an alarm system. If a thief tries to pick the device up, an acoustic warning signal sounds; the signal can only be deactivated by entering a personal code number.

**RoboMow,
Friendly Machines, Israel**

Automated Golf Caddie

When a professional golfer leisurely walks to the next tee, he calmly studies the surroundings and decides on a strategy for the next hole. Meanwhile, the caddie pulls along the clubs and bag behind him. The walk around the course soon becomes a hard grind for the hobby golfer, however, because he can rarely afford a caddie.

In the United States, golf is not just played by the elite; nearly every small town in North America has its own golf course. It is therefore no wonder that the GolfPro International company, based in Santa Clara, California, has invested several million dollars in developing an autonomous golf caddie robot. In the future, it will be rented out on American golf courses for about ten dollars a round.

The Intelecady is a computer-controlled, electrically powered golf caddie robot. Thanks to its telecommunication abilities and sensors, it can navigate around the golf course, follow a golfer, and perform the services that human caddies have performed until now. The electronic caddie must be able to recognize its golfer and follow him everywhere to do this.

Radio-Controlled Navigation Pinpoints the Player and the Environment

This is no easy task: The robot has to recognize the golfer's position relative to its own. It is also absolutely necessary that the robot be able to find its way around the golf course, and must therefore know its exact position. On top of that, it must be able to identify and avoid stationary and moving obstacles, such as trees, bushes, and other golfers, at the same time. The robot should also avoid the green and not fall into sand traps.

A small coded transmitter, carried in the golfer's pocket, takes care of personal identification; in this way, Intelecady follows only its player. A digital map of the golf course is stored on the intelligent robot's onboard computer. Because every golf course has areas where a robotic caddie should not go, these zones are marked separately on the digital map.

Steep inclines or downward slopes, small woods, and so on force the mobile robot to stop. Only when the golfer leaves this area does Intelecady start following him again. The robot goes into waiting mode in the area around the green so that the player is not disturbed by unnecessary movements while putting the ball.

The robot determines its actual position to within a meter on the map using a GPS (Global Positioning Satellite) system. If an even more precise navigation is required, such as when crossing a bridge, it uses white line markings on the path to find its bearings, which it identifies by way of optosensors.

The onboard computer receives the measurement values for accurate steering from encoders on both of the powered wheels. If unexpected obstacles such as other golfers or fallen branches crop up, eleven ultrasound sensors distributed around the circumference of the robot ensure safe operation.

In addition to carrying golf bags, Intelecady can supply the golfer with useful information. The player can ask it how far he is from the next hole, for example.

Along the way, however, the design engineers did not give the robotic caddie one thing: The machine does not supply the golfer with small talk and cannot comment on an unsuccessful stroke.

Intelecady, GolfPro Int., USA

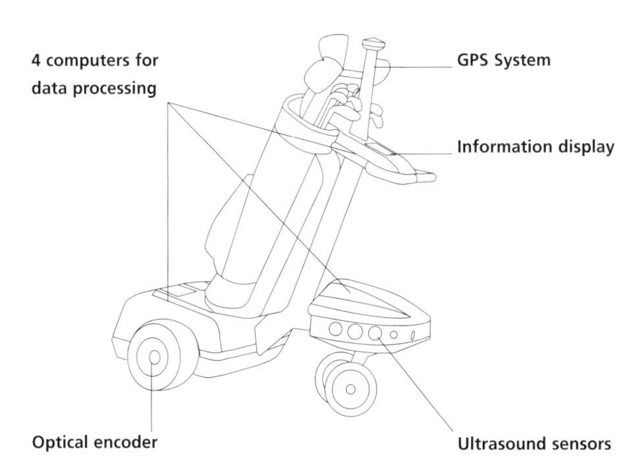

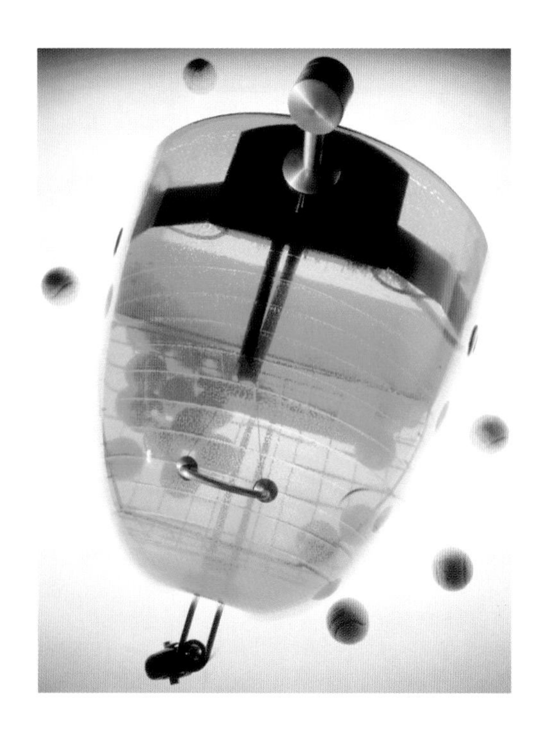

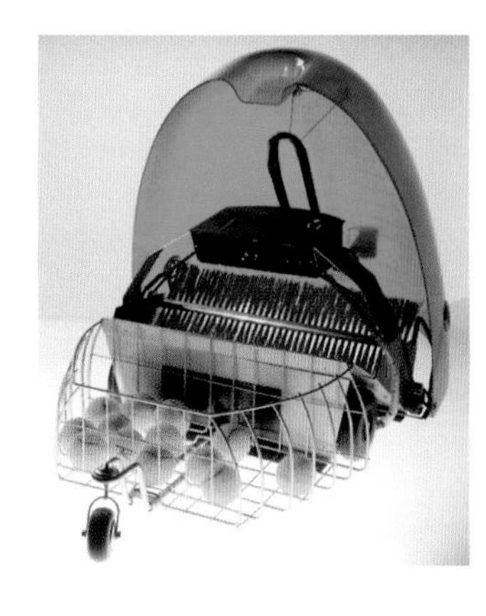

Inner workings of the robot

Protection Against Slipped Disks for Ball Boys

When Boris Becker is training, a lot of tennis balls end up lying on the ground. Who picks them up? This is a real problem for tennis schools!

In order to reduce the wear and tear on ball boys' backs and to promote robotic developments in this field, a competition was held by the American Association of Artificial Intelligence (AAAI) in August 1996 called "Clean Up the Tennis Court." Autonomous robots had to collect a maximum number of tennis balls in the shortest possible time and transport them in a container afterwards.

The winner, and thus world champion, was a tennis ball-collecting robot constructed by Hans Nopper in Stuttgart, Germany, together with Carnegie Mellon University (CMU) in Pittsburgh, Pennsylvania, and Real World Interface (RWI), in Jaffrey, New Hampshire. Professor Sebastian Thrun programmed the tennis ball collector, which had to travel on meandering loops around the court during the competition.

The autonomous vehicle is based on the mobile platform Pioneer 1, made by RWI. This robot, weighing just 9 kilograms, is made up of two powered wheels which are steered and adjusted separately and which are set in motion using DC motors. An additional passively steered wheel enhances stability. Five ultrasound distance sensors facing forwards and two facing backward ensure movement without collisions. The device, which costs about $2,800, is equipped with the software package Saphira, developed by SRI International, California, to assist sensor signal processing. This software also permits the recognition and extraction of features in the environment. Reactions to obstacles can also be programmed.

Brushes Instead of Bending Down

The actual system for gathering the balls consists of a CCD camera connected to Pioneer 1's onboard computer, a rotatable brush, and a collecting basket for the tennis balls. If a tennis ball is in the CCD camera's field of vision, the brush is then driven over a belt and the mobile robot steers toward the identified object. The rotating movement of the brush causes the ball to be hurled into the storage space at the rear of the robot. The collecting basket holds approximately 70 tennis balls. The cover, made of molded polycarbonate, can be tipped forward so that the balls can be removed.

Because of its high collecting capacity, this robot is pre-destined for use in tennis schools; tennis balls can even be collected during training with the ball cannon. The training breaks can be used instead for giving theory lessons, and the ball boys can go easy on their backs.

Tennis ball collector in action

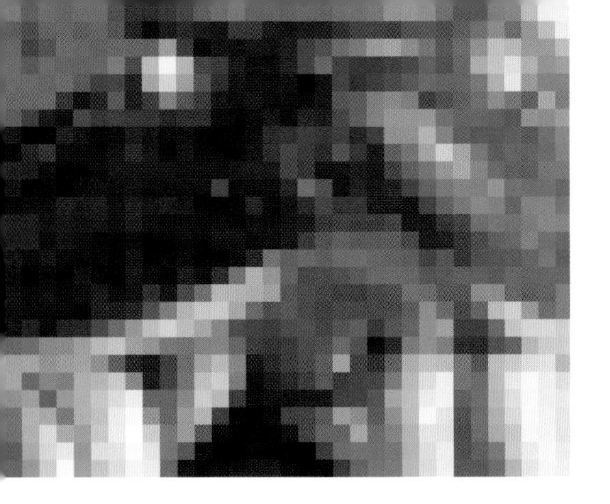

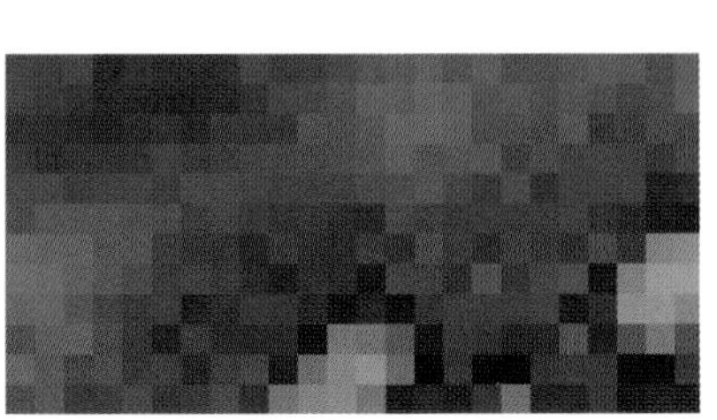

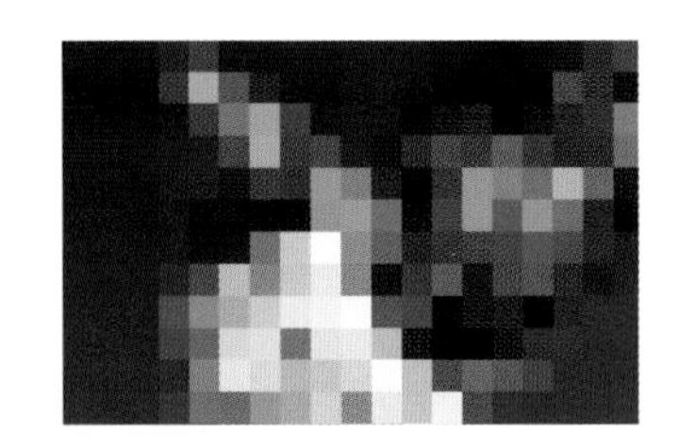

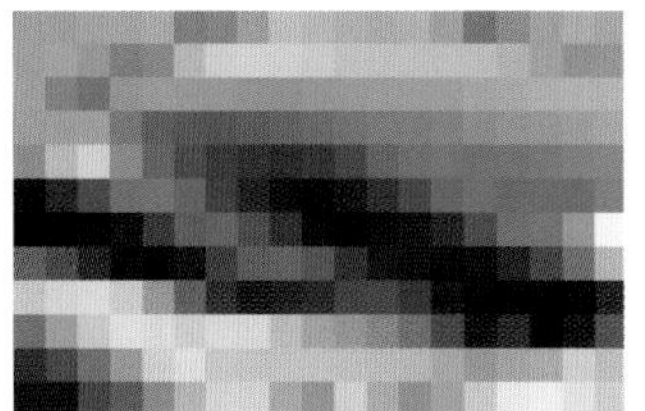

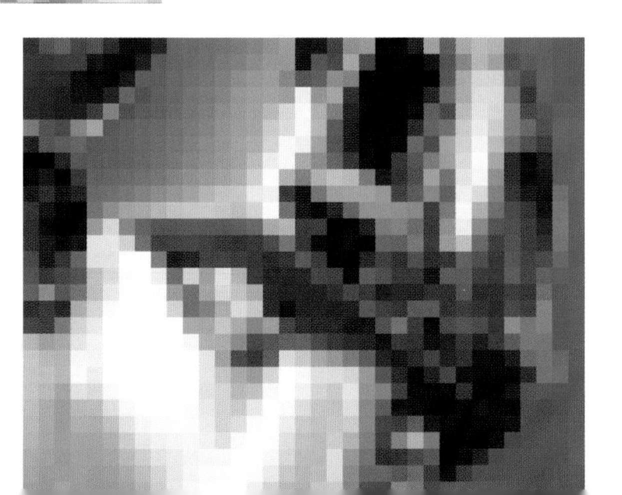

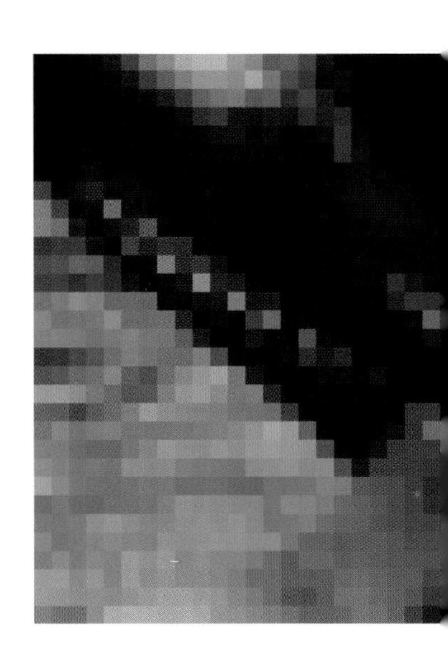

Entertainment

Killer Beasts with 56 Active Degrees of Freedom

Monsters have been a part of cinema history from its earliest days. They also make cinema history: King Kong, Godzilla, and Alien are ranked alongside Humphrey Bogart and Michelle Pfeiffer.

The film Abyss, as well as other virtual constructions from an SGI graphic workstation, ushered in the end of an era. No longer would furry animals with visible zippers mutate to become jungle monsters and giant space ships with the aid of camera perspectives.

Thanks to modern computer animation, special effects have become almost perfect illusions. Tom Hanks shakes John F. Kennedy's hand, and the audience is astounded. People are mistaken, however, if they think that puppets and models are going to disappear from the producer's bag of tricks because of computer animation.

The real superstars putting fear in viewer's hearts, and slowly but surely becoming more popular than the actors themselves, are deceptively realistic reproductions of monsters, in whose veins hydraulic oil rather than blood flows. Their computer brains have the power of a mainframe, and their bones are made of high-grade steel, aluminum alloys, and the most up-to-date compound materials.

Animatronics Places High Demands on Robotics

These highly developed animatronics substitutes, which aeronautics and robotics engineers are raving about because of their complexity, can only come into existence where money is of no importance and design engineers are only concerned about the time factor. America's film producers swear by the entertainment robots made by the two California companies Edge Innovations and Stan Winston Studios.

Not only in Hollywood do people like to be entertained by robots, however. Sarcos Entertainment Systems (SES), based in Salt Lake City, Utah, manufactures robotic systems that attract visitors in droves at trade fairs and in leisure parks. Even the Massachusetts Institute of Technology (MIT) in Boston, Massachusetts, is devoting research work to constructing a two-legged, freely moving dinosaur.

Original or fake?
Robotic shark (left) and robotic whale Willy (right), Edge Innovations, USA

Anaconda with its master Walt Conti,
Edge Innovations, USA
(Photo: pixel&zeichen, Hamburg, Germany)

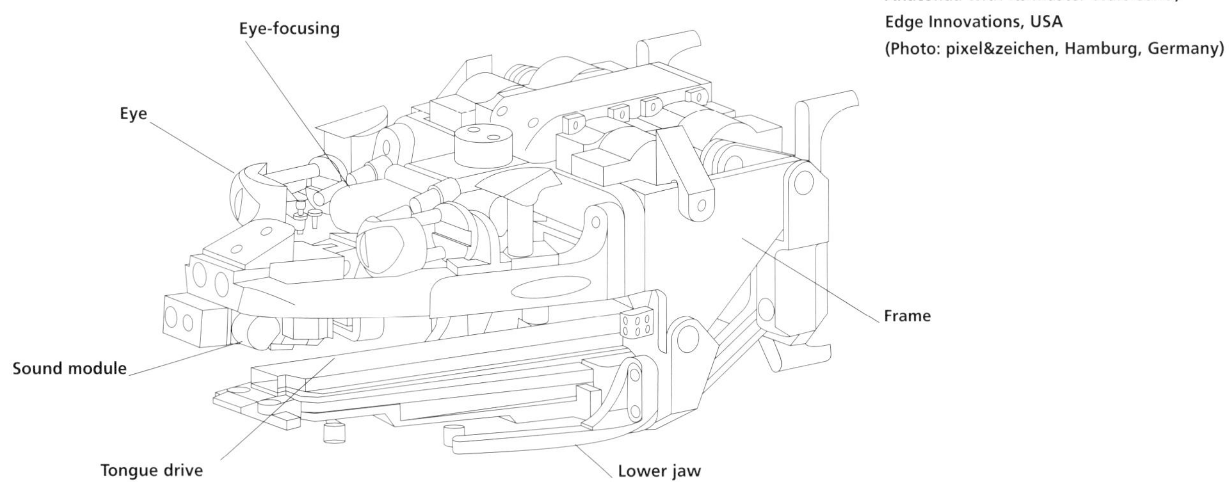

Hollywood's Killer Beasts Are Exclusive Entertainment Robots

In Columbia Tristar's thriller, Anaconda, a monster decimates the members of a research team staying in the Amazon area. The star of this film is a twelve-meter-long giant snake weighing several tons. It was developed within a few months by Edge Innovations.

The fully functional replica of the constricting snake is composed of 60 artificial vertebrae driven by hydraulic cylinders, each consisting of 250 individual parts made of high-tensile steel. Other materials could hardly withstand the snake's weight or the enormous accelerating forces of the remarkably dynamic colossus: The monster's head whips through the air at up to 60 kilometers per hour.

The work stations for the motion simulation and control of this entertainment robot have a computational power of around 50 PCs. The snake's body, which is 30 centimeters thick, contains 60 kilometers of electrical wiring and hydraulic cables wound up inside it. In order to move the killer beast, a hydraulic aggregate powered by several hundred horsepower pumps oil into the artificial arteries of the colossal creature. Special control concepts had to be developed and implemented for the anaconda because neither industrial robots nor aeronautic manipulators with such high degrees of freedom exist.

Stan Winston, Edge Innovations' biggest competitor, lives near the large film studios of Los Angeles. Man-eating science fiction monsters such as Alien and anthropomorphic robots such as Terminator are made in his labs.

The stars of "Deep Blue Sea"- robotic sharks, Edge Innovations

Working on the film Anaconda, in the jungle and at the Edge Innovations labs

Master-slave robot, Sarcos Entertainment Systems (SES), USA

Puppeteers Make the Master-Slave Robot Dance

The movements of entertainment robots are usually controlled using a joystick or master manipulator. Puppeteers turn master-slave robots into interactive actors that react in real-time. Depending on the complexity of the movement, several people may be responsible for the various parts of the robot's body. The dinosaurs in Steven Spielberg's Jurassic Park and The Lost World were sometimes operated by up to six puppeteers: Face, head, neck, torso, and limbs each had their own master.

Sarcos Entertainment Systems (SES) develops and manufactures anthropomorphic robotic systems that can be programmed or controlled using a sensor suit. A sensor suit for controlling anthropomorphic robots can be operated by one person. This is all the more surprising because man-like robots possess up to 56 degrees of freedom which, in extreme cases, may all need to be directed simultaneously.

A range of sensors have been built into the suit, which registers the master's movements and transfers them to the robot's control system. The variable algorithms, developed for this type of application by SES, in collaboration with the University of Utah's Center for Engineering Design, ensure that the machine's movements resemble those of man more closely than ever. The sensor suit enables real-time interaction between the anthropomorphic robot and the audience.

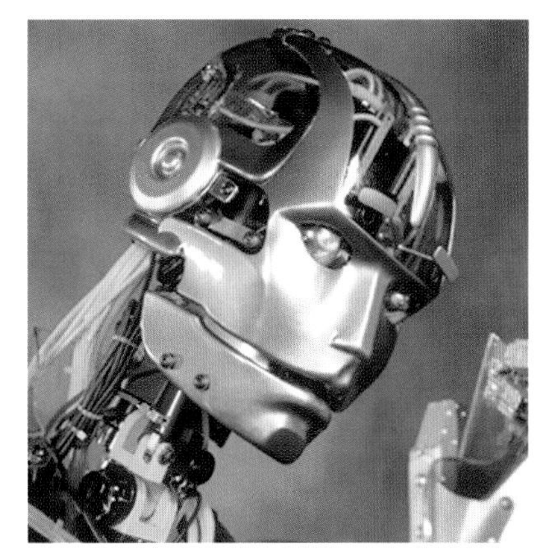

**Entertainer robot's head,
Sarcos Entertainment Systems (SES), USA**

Dinosaurs - Completely Detached

The whale made by Edge Innovations for Free Willy had an umbilical cord over 90 meters long, which supplied hydraulic oil and electricity, that had to be retouched afterwards so that it was not visible on film. To avoid this problem in the future, scientists from MIT are working at present on a two-legged robot that will be able to move autonomously.

Walking robot, MIT Leg Laboratory

The construction of the therapod dinosaur's body is an excellent example of an agile, quick-moving, two-legged robot. Since 1996, Peter Dilworth has been working at MIT's Leg Laboratory on the first autonomous, dynamically balanced, two-legged walking robot, which carries its batteries and control computer with it.

While the control problems associated with the gait of the two-legged robot have been solved for the most part, the mechanical design of this research prototype's motional apparatus has become the most important consideration. Miniaturized electromotors and spring elements should give the artificial dinosaur unprecedented ways of moving.

Although it was declared dead years ago, the field of animatronics is experiencing a boom these days that should last indefinitely. After aeronautics, animatronics is blazing the trail for future innovative technologies.

Peter Dilworth (right) and therapod dinosaur robot, MIT Leg Laboratory, USA

Walking therapod dinosaur robot,
MIT Leg Laboratory, USA

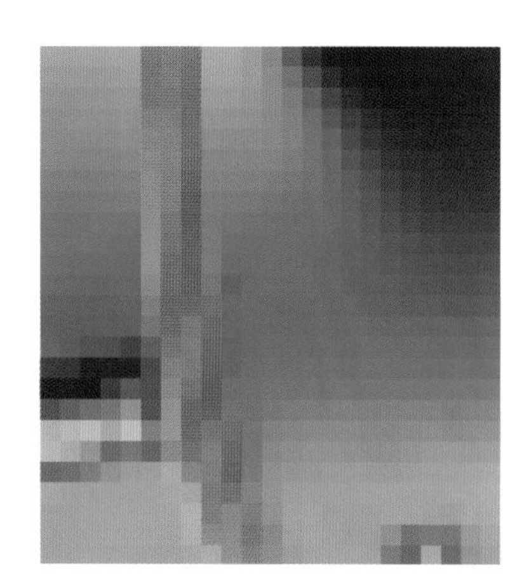

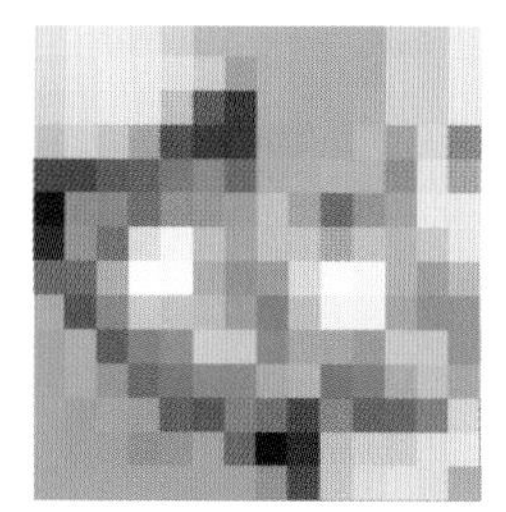

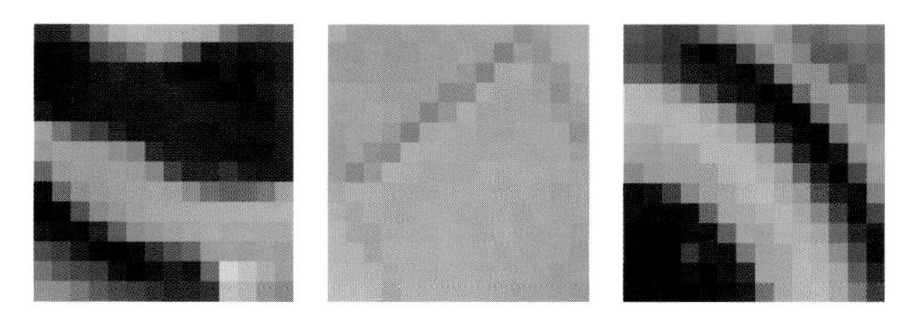

Nursing care

More Time for Care

Guaranteeing humane living conditions for both the elderly and for people dependent on care and assistance is fundamental to a responsible society.

In Germany alone, the number of 60-year-old citizens will have doubled and the number of 90-year-olds will have tripled by the year 2030. As a result, the number of people requiring care will also increase. By the year 2040, it is estimated that 3.5% of the population will need care; today, this figure is only 2.1%.

An important objective for those needing help is an ability to participate in family and social life. Following this principle of "home care instead of hospital care" will become a challenging requirement in the near future.

The physical strain on individual care givers is high, making it a burden on the affected individuals and on the cost of nursing care. A high priority is given to finding ways to curb costs in both an effective and ethically acceptable way while still maintaining patients' quality-of-life. Personnel are scarce for jobs which, hopefully, will never be carried out by machines: providing care and support for unique individuals.

Home Care Systems: Rather At Home Than in a Nursing Home

Considerable sums could be saved on individual care in retirement or nursing homes if those people requiring care and assistance could provide for themselves longer at home. Home care systems will assume a key role in implementing this concept.

The term "home care systems" denotes technical assistance systems for people dependent on care and assistance in the home. Preliminary solutions already exist for using service robots as intelligent aids. A fundamental differentiation is made among three steps in the development of technical systems.

In Step 1, manipulators were developed to be mounted on tables or benches. These robotic arms, such as the Devar System made by the Tolfa Corporation in Palo Alto, California, or the Raid System manufactured by Oxford Intelligent Machines Ltd. in England, are particularly suitable for people who, due to a handicap, require assistance to carry out their occupation. Such solutions are found only rarely in the home.

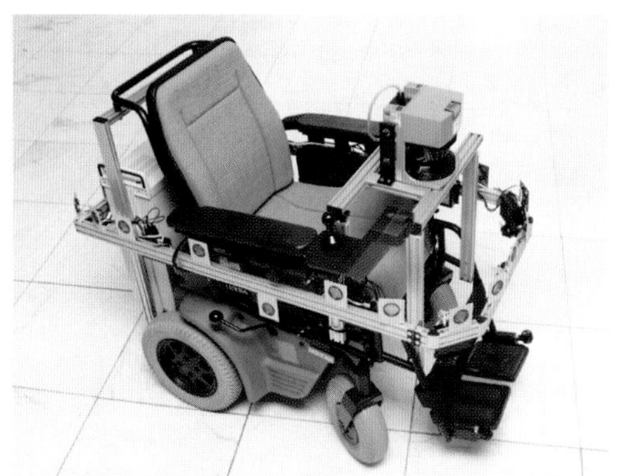

Maid, FAW Ulm, Germany

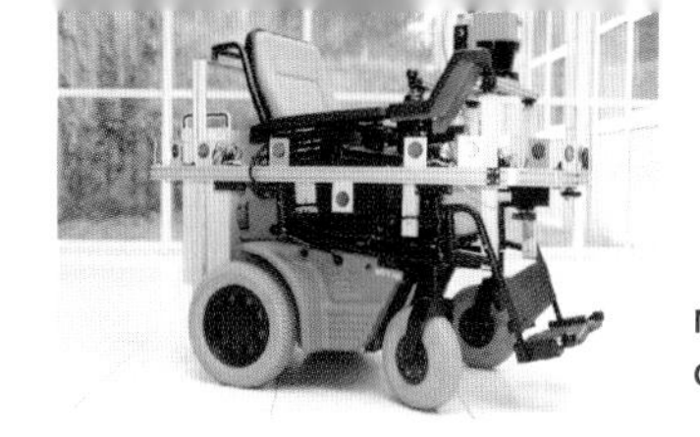

MAid, FAW Ulm, Germany

The Future Belongs to Mobile Multipurpose Robots

In Step 2, more interest was taken in robotic arms that could also be mounted on wheelchairs. This step, however, only helps people who are confined to a wheelchair as a result of their handicap.

A much wider spectrum of application is covered by partially or fully autonomous mobile robots. These are the prototypes resulting from Step 3 in development. They can perform an almost inexhaustible range of duties in the home. This advancement equally affects those physically challenged, those confined to bed, and elderly persons who no longer have the physical energy or agility to perform various tasks at home.

Robotic Arm for Wheelchairs

The Dutch company Exact Dynamics BV, in Zevenaar, has developed Manus, a robotic arm for wheelchairs. Manus helps handicapped people to warm food in a microwave oven, brew tea or coffee, open doors, or use such items as an electric toothbrush.

Manus can be attached either to the right- or left-hand side of a suitable wheelchair. Manus has six rotary joints, a prismatic joint at the base, and a gripper. Sliding clutches in the power train provide additional security. The mechanical arm is controlled by using a joystick or keyboard. Manus can handle a payload of 2 kilograms with its gripping arm, which is 85 centimeters long when fully extended; the device itself weighs 20 kilograms.

More Mobility for the Severely Handicapped

MAid is an intelligent, self-driving wheelchair for people with particularly poor muscular function. Handicapped people suffering from multiple sclerosis will benefit from it, as well as people who are paralyzed or who are afflicted with muscular diseases. MAid can function in either partially or fully autonomous operational mode.

With the partially autonomous mode of operation, the user is able to initiate individual, spatially limited driving maneuvers, such as reversing into the bathroom or driving through a narrow doorway. In fully autonomous operational mode, MAid carries out driving maneuvers, including more distant movements, without interacting with the user. MAid can travel accurately and without collisions, even in a busy shopping mall or train station.

The mechanism is based on the Sprinter wheelchair, manufactured by the Meyra company, which is equipped with a differential drive and castor wheels. Connected to an industrial PC capable of real-time processing, the sensor system consists of 22 Polaroid ultrasound sensors, a fiber optic gyroscope, angle transducers on the drive wheels, several infrared scanners, and a two-dimensional laser scanner.

MAid has been developed at the FAW in Ulm, Germany, within the scope of the cooperative project Inservum (Intelligent Service Environments), sponsored by the German Federal Ministry of Education, Science, Research and Technology (BMBF). The companies Erlau AG, reis robotics, Robert Bosch GmbH, Siemens AG, Forwiss, Fraunhofer IPA, and TÜV Südwest were partners in the project.

Robotic arm "Manus", Exact Dynamics BV, Holland

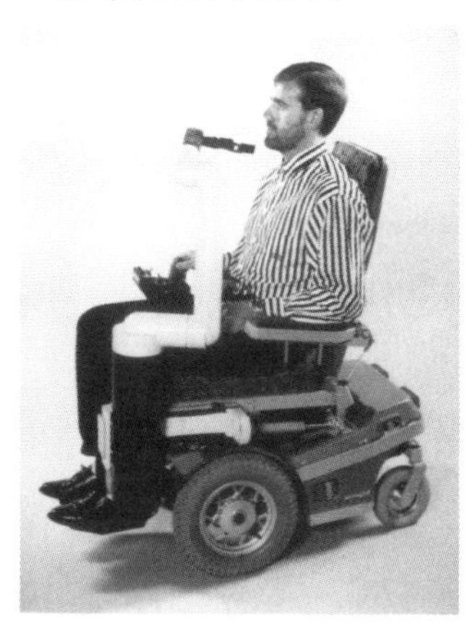

castor wheels
Castor wheels are rotatable, passive stabilizing wheels

Mobile unit of the Urmad system

Mobile Helper in Nursing Homes

The Italian-made Urmad system is a mobile unit that can navigate autonomously in a known environment, maneuver around obstacles, and pick up and transport objects. The Urmad system was used to research the potential of implementing a mobile, autonomous assistant in nursing homes. The system, which only exists as a prototype at this time, is comprised of a stationary and a mobile subsystem.

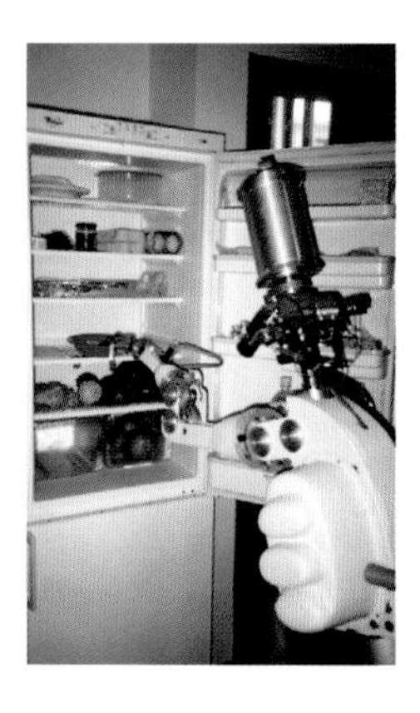

Movaid – interaction with adapted kitchen appliances

The stationary unit of the Urmad system is within reach of the user and forms the man-machine interface. It consists of a communication unit to assure data transfer between the stationary unit and the mobile one, and a PC with a graphical user interface.

The mobile Urmad unit, a three-wheeled vehicle, functions as the transport system for a robotic arm that has eight degrees of freedom and which has been specially adapted for use in the home. A three-fingered hand possessing two degrees of freedom is situated at the extremity of the robotic arm; the fingertips and the palm of the hand are equipped with force and tactile sensors. The mobile unit has a camera system for navigation and object identification purposes. Ultrasound distance sensors, arranged in a ring-shape, are responsible for collision avoidance.

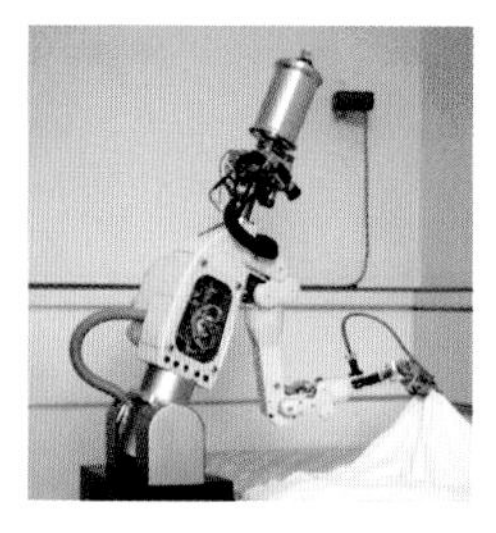

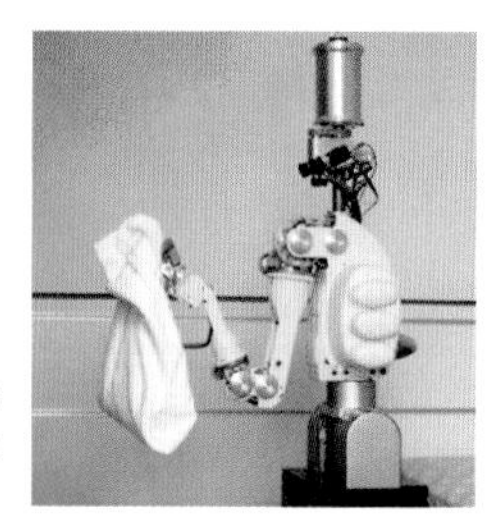
Movaid working as the chambermaid

The Urmad system was developed between 1992 and 1994 by an Italian consortium consisting of ten partners, most of which were universities. The participants were the Scuola Superiore S. Anna in Pisa; the universities of Genoa, Naples, Florence, and Bologna; the Milan Polytechnic College; the Telerobot Consortium in Genoa; Scienzia Machinale srl. in Pisa; Aitek srl. in Genoa; and the Centro Protesi INAIL in Bologna.

The Physically Challenged Can Also Continue to Be Independent in Their Own Homes

Because of the high energy requirements of its components, its weight, and its dimensions, which are unsuitable for use in the home, it has been possible to implement the Urmad system only in hospitals and retirement homes having a suitable infrastructure (elevators, wide corridors, and so on). A prototype called Movaid has been developed by a European consortium for use in the home.

Movaid (Mobility and activity assistance systems for the disabled) is a distributed robotic system. The semi-autonomous mobile robotic unit is able to dock at several stationary communication units. A wide range of standard kitchen equipment also forms part of the system, and the operating interfaces of these devices have been specially adapted to meet the needs of elderly or physically challenged people. Items such as a microwave oven or lemon squeezer have been redesigned so that they can be operated more easily by physically challenged people.

Movaid's stationary communication units have a multimedia man-machine interface derived from the Urmad system, which enables easy interaction between the operator and the robotic system.

Movaid's mobile robotic unit is a transport system with two steerable and two driven wheels, on which the robotic arm used in the Urmad prototype is placed. This mobile segment also has ultrasound distance sensors arranged in a circular shape, and a camera system for navigation and object identification.

The modular design of the robotic unit allows for the assembly of the robotic arm on other transport systems, such as wheelchairs. It also enables the drive unit to be used as a mobile platform for other purposes. Compared with the Urmad system, it has been possible to reduce Movaid's dimensions so significantly that this unit can now be used in the home.

Coordinated by the Scuola Superiore S. Anna in Pisa, Italy, the Movaid project has been carried out by an international consortium with the participation of Philips CD, Holland; S.M. Scienza Machinale srl., Italy; Domus Academy, Italy; Biotrast, Greece; Inserm Unité 103, France; FST, Switzerland; the Italian universities of Ancona and Genoa; and the Commisariat Energie Atomique, France.

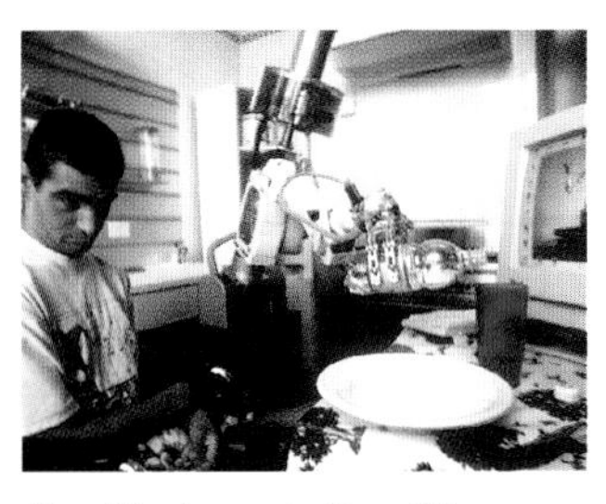

Movaid - demonstration of its precision movement

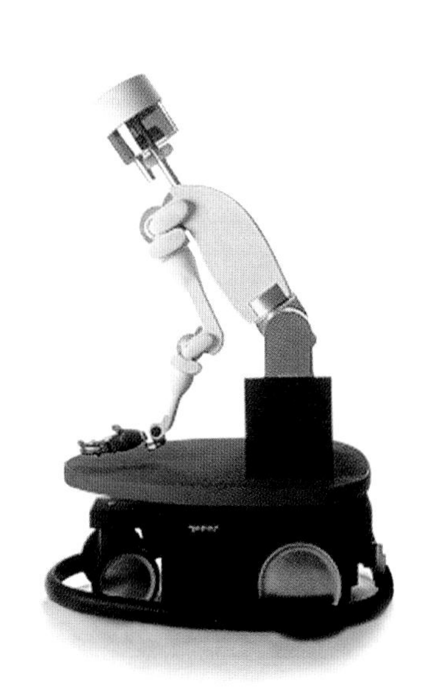

The Care-O-bot Obeys Spoken Commands

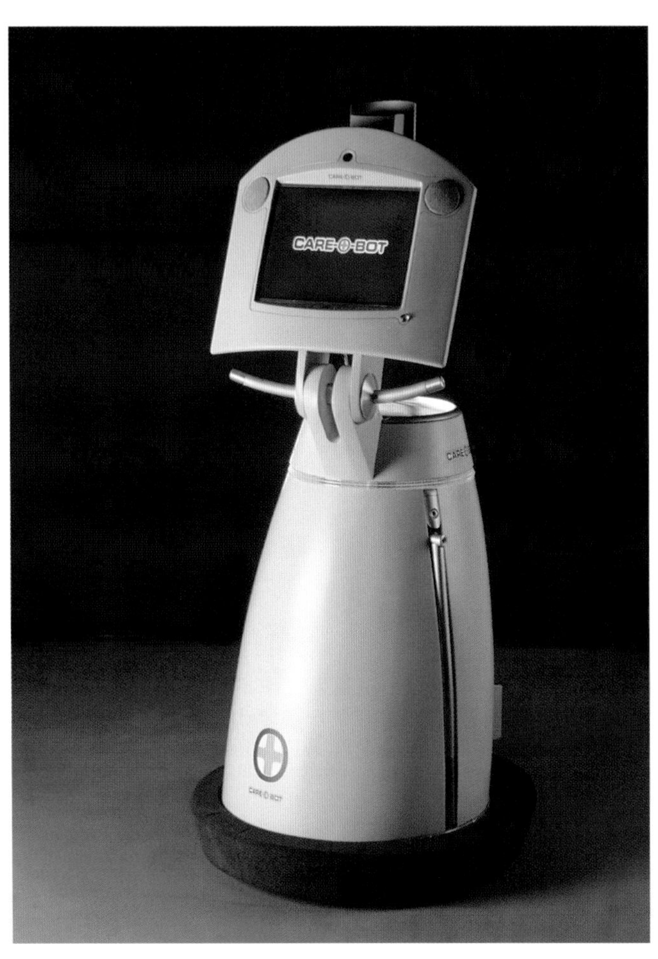

Care-O-bot, Fraunhofer IPA, Germany
Designed by Katja Severin and
Hans Nopper, Stuttgart, Germany

Care-O-bot has been developed at Fraunhofer IPA as a futuristic home care system. It enables elderly people and those dependent on care or assistance to stay in their home environment longer.

The independent, mobile service robot Care-O-bot responds to voice commands. It can directly assist dependent people and relieve the burden on nurses, because it provides help in the most important aspects of everyday life:

- Communicating with public authorities, doctors, and so on; personal communication; daytime manager, media management; distress call
- Providing food and drinks, supply and disposal, cleaning tasks
- Controlling household infrastructure, such as heating, lighting, alarm system, and so on
- Handling assistance using gripping, lifting, stopping, and provision aids; help in dressing
- Assisting mobility by way of supporting aids and walking aids and helping a person to stand
- Guarding personal safety by monitoring vital functions and, when necessary, setting off an emergency alarm

To perform these tasks, the Care-O-bot prototype has been designed as a mobile autonomous platform. Two arms affixed at its sides, each with one degree of freedom, function as active standing and walking aids. The integration of a robotic arm with several joints is presently being planned.

Weighing 140 kilograms and powered by two brushless servo motors via a differential drive, the Care-O-bot can travel up to 1.5 meters per

second on its two solid rubber driving wheels and four stabilizing wheels. The 46 Ah (ampere-hour) batteries can be recharged in just 10 minutes.

Angle transducers attached to the wheels for odometry, a laser scanner, and a stereo camera record the Care-O-bot's environment. It can be navigated by the operator by using both spoken commands and a touchscreen; the system either makes spoken announcements or displays them on a 14-inch TFT screen. The mobile unit, networked with CAN and Ethernet, is linked via a radio LAN to the central computer.

Examples of the Care-O-bot's range of functions

CAN
Controller-Area Networks, see Chapter 20.

LAN
Local-Area Network. Networking several computer systems.

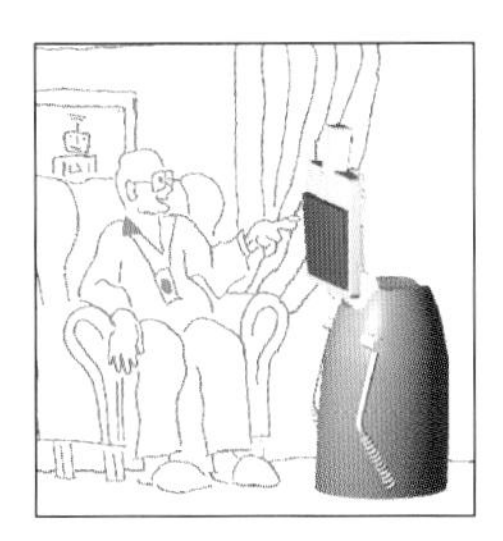

Simple Handling Is a Top Priority With Care Robots

The multi-media man-machine interface consists of an audiovisual and a tactile communication unit with a microphone, a loud speaker and a touchscreen. For navigation and object identification purposes, the Care-O-bot is equipped with a two-dimensional laser scanner, several CCD camera systems and ultrasound distance sensors.

Over the next few years, further developments will be made in the complete system and sub-systems for use in nursing care. Service robots for nursing and care are used in areas with a low level of technology, and they are used and operated by untrained people. Therefore, current research focuses on how the robots will interact with people.

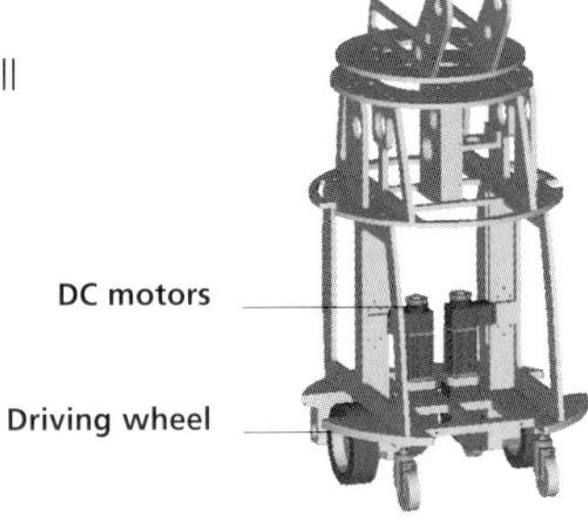

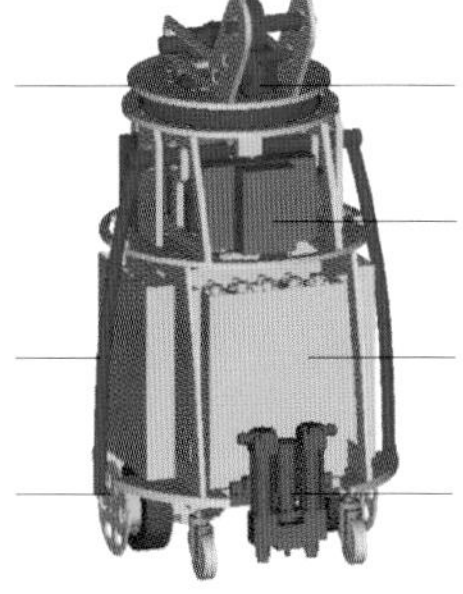

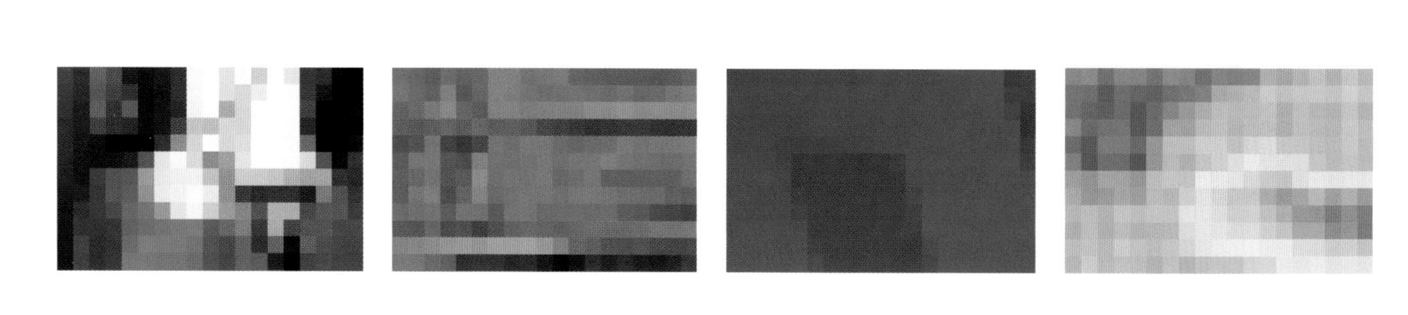

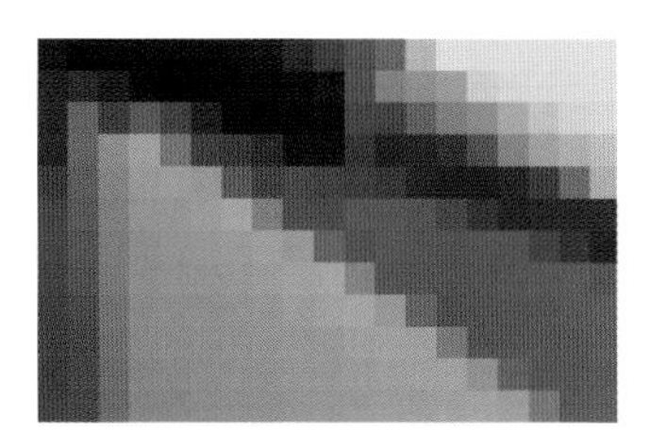

Medicine

Manipulator
HANDY, Japan

The Surgeon's Third Arm

Few fields need qualified people as much as the field of medicine. No one even considers replacing doctors or trained nurses with robots. They would be able to perform their important jobs much better, however, if we could reduce their physically tiring tasks to a minimum and provide optimal assistance for the routine work and more difficult tasks they have to perform. For this reason, scientists and engineers all over the world are developing robot-assisted systems which are either able to carry out individual handling tasks in hospitals independently or able to provide technical support.

The term "medical robot" has not been adequately defined and refers to many different applications, ranging from the simplest laboratory automation to the highly complex surgical robot. In the field of medicine, the classic service robot is fundamentally different from passive robotic systems and even more so from those used in surgical situations.

Medical service robots include laboratory robots, which execute hundreds of lab tests simultaneously while carrying out their work quickly, accurately, and tirelessly. Mobile robots for distributing drugs or meals in hospitals are also included, as well as robots for helping to lift patients or assist handicapped people to eat, read, and so on.

Intelligent Robots in the Operating Room

During procedures, robot-assisted systems can relieve the surgeon of physically and psychologically tiring tasks. This will allow him to focus more on his actual medical tasks – diagnosis and treatment – while still maintaining the responsibility for patient safety and for the success of the surgery.

Intelligent robotic systems can assist in three types of surgical tasks:

- Long procedures, by holding instruments or handling them for extended periods of time. These are arduous tasks, which deplete the surgeon's ability to concentrate and which often must be performed in a tiring physical posture.
- Laborious, handyman-type procedures such as preparing a femur for the implantation of an artificial joint. A robot works with a very high level of mechanical precision which would be unattainable by hand.
- Surgical operations on micro-structures, which would be impossible to perform without the technical assistance of actuators and surgical microscopes.

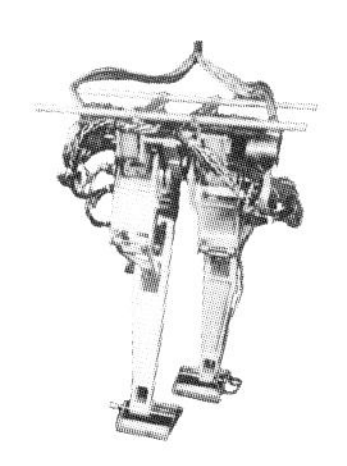

Walking aid BIPED, MITI, Japan

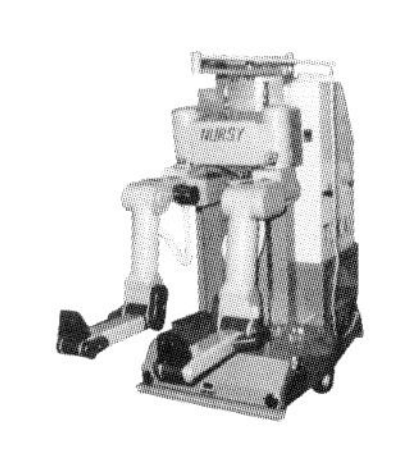

Nursy, the Tokay company, Japan

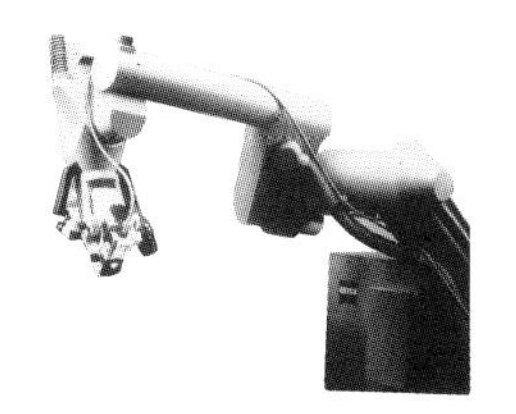

Multiple coordinate manipulator MKM, Zeiss, Germany

The more simple robots are passive systems which function as extended instrument holders, but with which no operating measures are carried out.

Surgical robotic systems offer a much higher level of functionality; a differentiation is made below between autonomous and guided systems.

- Autonomous surgical robotic systems perform individual program-controlled steps during an operation.
- Guided surgical robotic systems bridge the gap between a doctor's diagnostic ability and his fine motor limitations. It is true that many diseases can be diagnosed now using affectors such as microscopes, but the instruments (effectors) are not available to carry out precise operations in the sub-millimeter range, such as neurological microstructures. Robot-assisted systems can convert the normal movements made by the hands of a surgeon into the smaller and, therefore, much finer movements of an instrument.

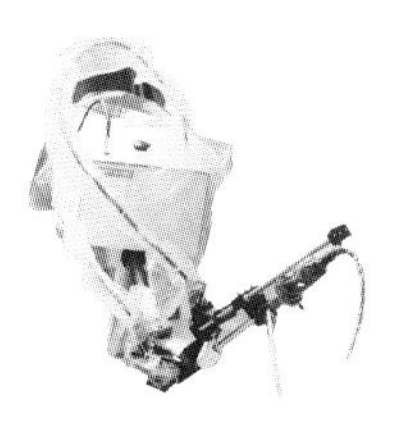

LAPAROBOT, Armstrong Projects PLC, USA

laparoscope
A laparascope is a specially designed, rigid endoscope that is introduced into the abdominal cavity through a very small incision in the region of the umbilicus. The abdominal wall is then elevated using air or CO_2. This allows all the abdominal and pelvic organs to be examined and, if necessary, to be drained or biopsied.
A local anesthetic is used in the procedure.

minimally invasive surgery
The term "minimally invasive surgery" includes surgical, diagnostic, and therapeutic techniques which, as opposed to conventional surgical techniques, allows the operation to be carried out on diseased or damaged tissue either through a natural orifice in the body, such as the urethra, or through an artificially-created opening (penny-sized incision).

endoscope
An endoscope is an optical instrument attached to a light source for examining the interior of hollow organs and body cavities and for taking tissue samples.

Many Robots Are Already in Daily Use

The classic service robot, HelpMate, manufactured by TRC (Transitions Research Corporation), has already been implemented in about 20 hospitals throughout America and Japan. It is able to transport meals, bed linens, drugs, and many other items within a hospital, and thus reduces the strain of routine tasks on hospital employees.

Benign prostate hyperplasia (a non-malignant swelling of the prostate gland) is the most common cause of bladder problems in elderly men. A transurethral resection (TUR), i.e. the partial removal of the prostate via the ureter, is a proven minimally invasive treatment. At the Imperial College in London, robot-assisted TUR of the prostate has been tested. The first experiments were performed using LARS, a Puma 560 robot.

Telemanipulators Take the Strain off Guiding Endoscopes

AESOP stands for Automated Endoscopic System for Optimal Positioning. This system guides and positions a laparoscope for abdominal investigations in minimally invasive surgery. The endoscopic carrier system AESOP 2000 enables the endoscope to be guided by speech. The surgeon can keep his eyes and hands fully focused on the actual operating procedure while directing the endoscope using spoken instructions.

This surgical robotic arm, developed by Computer Motion, in Goleta, California, has already successfully assisted in over 30,000 minimally invasive operations and has been installed in over 300 clinics and surgical practices.

ARTEMIS, a system developed at the Forschungszentrum in Karlsruhe, Germany, makes up a part of the carrier system. It is an advanced robotic and telemanipulator system with which the surgeon can perform minimally invasive abdominal operations from a workstation. ARTEMIS consists of two different telemanipulation working units: Tiska, a computer-controlled carrier system for surgical effectors, and Robox, a computer-controlled endoscopic guiding system.

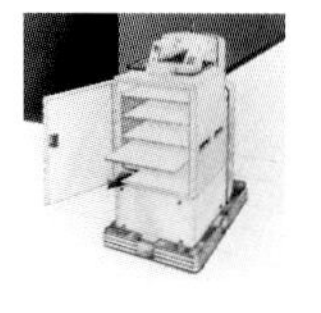

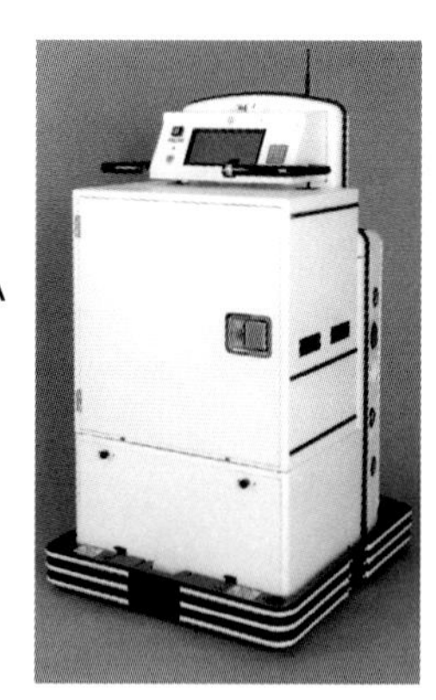

HELPMATE, TRC, USA

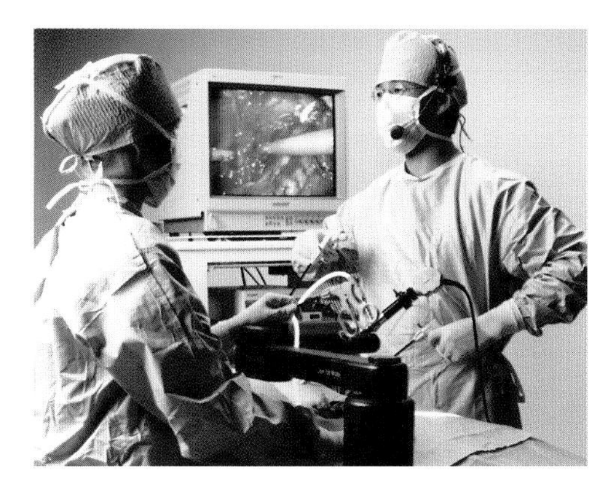

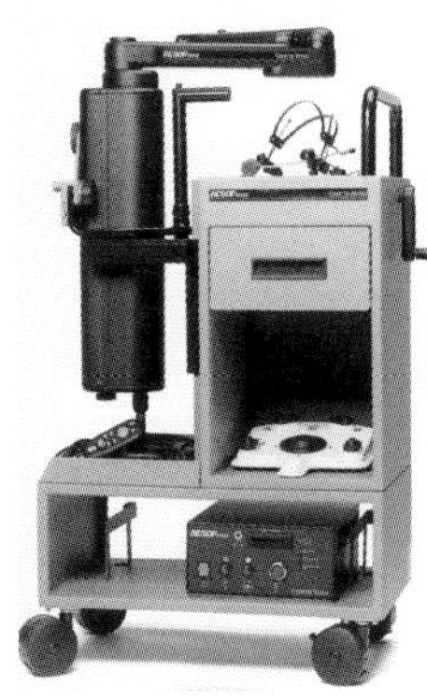

Endoscopic carrier system AESOP 2000, Computer Motion, USA

ARTEMIS, Forschungszentrum Karlsruhe, Germany

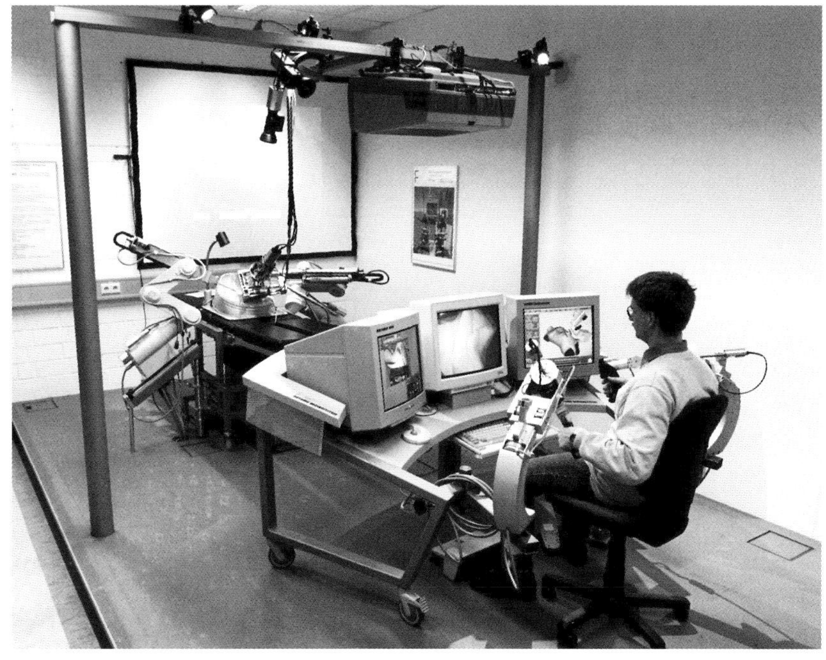

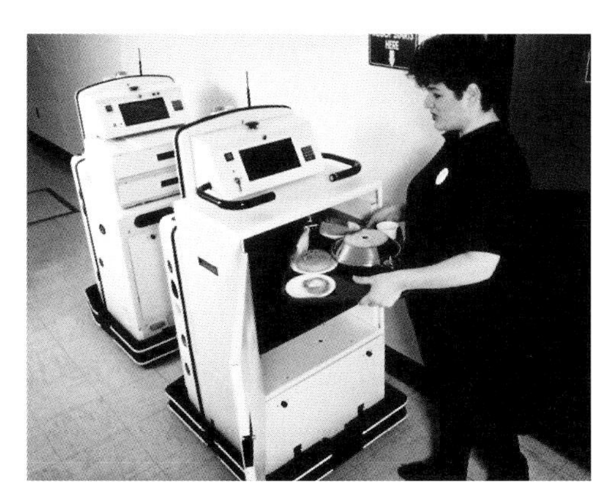

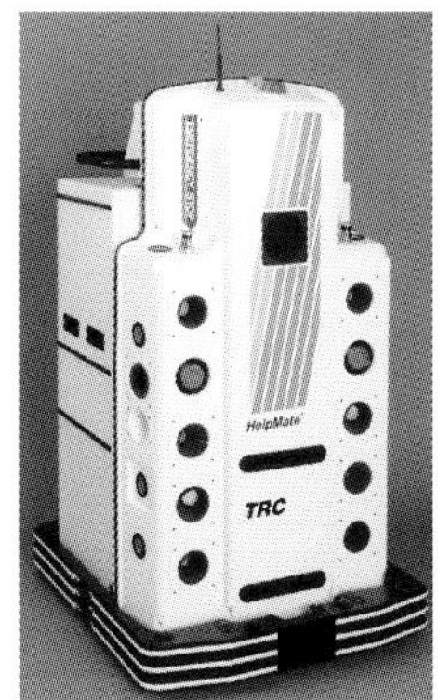

HELPMATE, TRC, USA

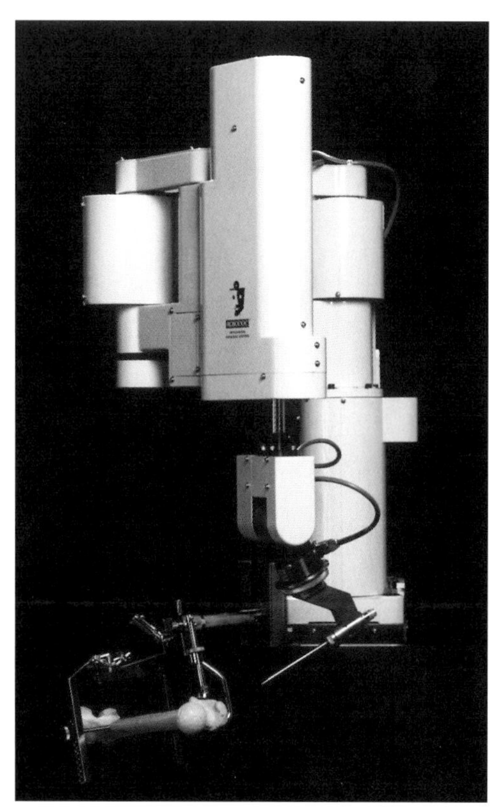

Surgical robot
ROBODOC, ISS, USA

Increased Precision for Joint Prosthetics

A fundamental problem in replacing "large" joints, such as hips, with artificial endoprostheses is the exact fitting of the prosthetic shaft into the bone. Improving the contact between bone and prosthesis improves the probability of acceptance of the prosthesis by the bone. Only increased precision of the secure fit of the artificial hip joint will allow the joint to endure mechanical strain and to last for several decades.

In clinics all over the world, more than 900 hip operations have been performed using ROBODOC, a robot manufactured by Integrated Surgical Systems (ISS) in Sacramento, California. ROBODOC can prepare the channel for the prosthesis (cavity) from the shaft and socket much more accurately than the best surgeon could. Before the cement-free implantation of an endoprosthesis is performed, a precise three-dimensional plan is created using the ORTHODOC graphical computer. Here, with the aid of the computer tomogram of the femur, the endoprosthesis with the best fit is selected. During planning, a precision alignment of ±0.1 millimeter and ±0.1° is attained in the bone marrow cavity. The precision of the robotic milling is ±0.5 millimeter.

computer tomography
A computer tomograph scanner (CT) takes X-ray images of the patient from various angles, stores them, and obtains plane sections from them, giving three-dimensional structural information about the patient's internal organs and cavities, which can only be seen using this method.

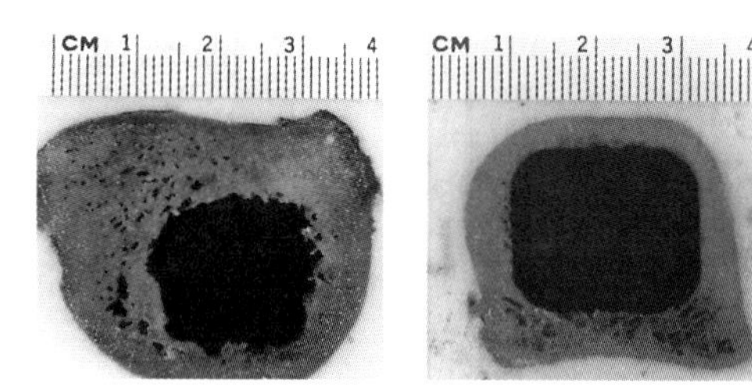

Comparison of cross-sections of bone: on the left, a manually prepared bone; on the right, a robot-assisted prepared bone

Preoperative planning workstation,
ORTHODOC, ISS, USA

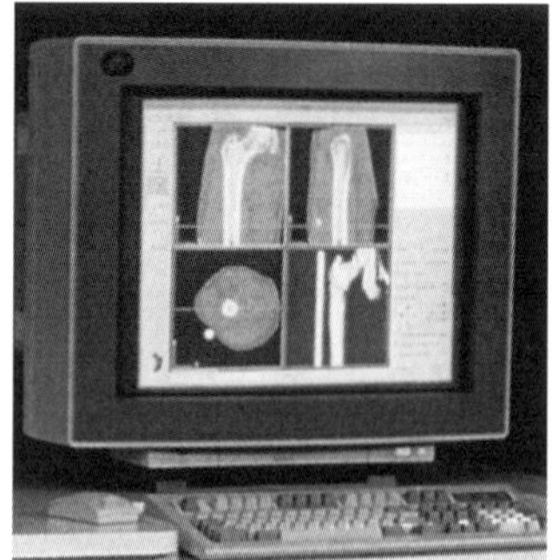

The CASPAR system (Computer Assisted Surgical Planning and Robotics) made by ortoMaquet is also preparing cavities for human hip endoprostheses. The pre-operative planning tool has been integrated into the Proton Station and selects the correct endoprosthesis based on three-dimensional data. The NC data for the manipulator is also calculated here, using the finite element method (FEM) to check the mechanical connection between the bone and the shaft of the prosthesis. The robot receives the planning data and mills the planned shaft profile in the bone with a high level of precision. The robot's level of precision is within ±0.5 millimeters with a deviation from the shape of less than ±0.2 millimeters.

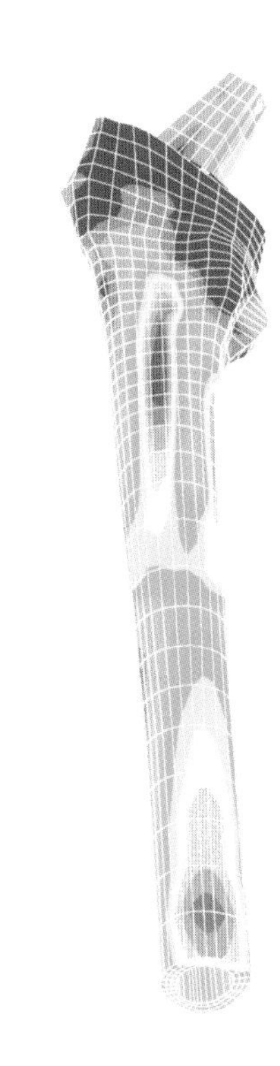

Distribution of tension for analyzing the fit of the endoprosthesis using finite elements

NC data

Numerical control (NC) data means the description of a robot's path using discrete coordinates that can also be ascertained using a manually guided teach-in process

FEM

FEM stands for finite element method. It is a numerical process whereby the object to be described or analyzed is divided into a finite number of elements.

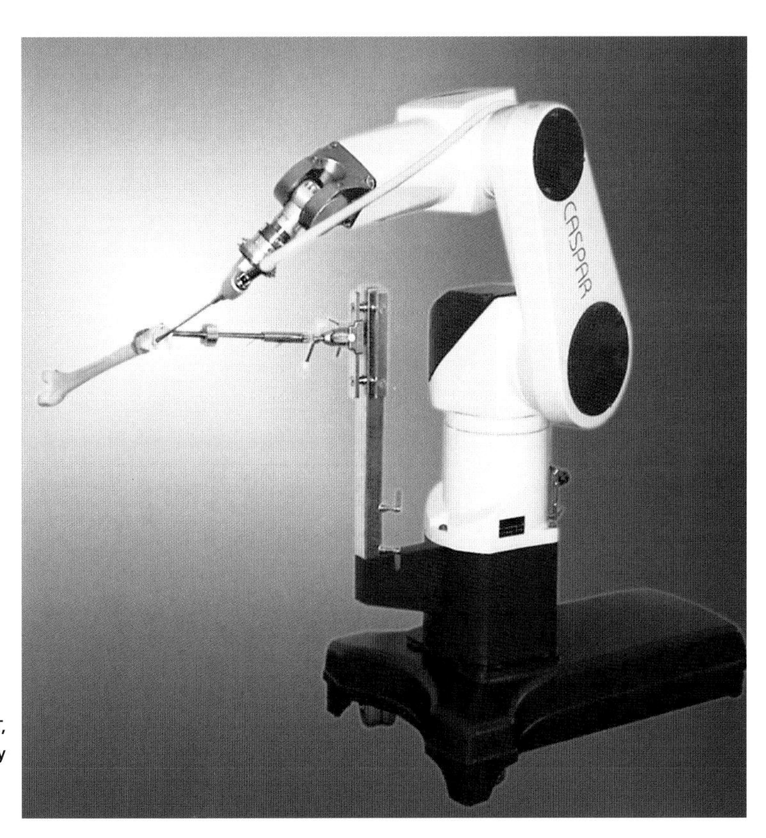

CASPAR, ortoMAQUET, Germany

stereotaxis

Stereotactic operations are procedures in which deep brain structures can be precisely located and operated on with the aid of a framework system and a thin probe inserted through a small drill hole in the skull

Extreme Accuracy in Neurosurgery

Stereotactic operations are among the most difficult in the field of neurosurgery because specific areas of the brain must be reached and operated on with absolute accuracy. A fraction of a millimeter can decide the success or failure of such an operation.

The neurosurgical robot MINERVA performs conventional stereotactic operations while the patient is lying inside the computer tomography scanner (CT). This allows the surgeon to follow the position of the instruments with a real-time image of the operating field. Because of online imaging, the visualization software and the precision of the robotic system, a whole range of operations can be performed on the brain, such as biopsies, stimulation, or the implantation of electrodes.

The program-controlled robot makes an incision in the scalp, drills through the skull, and supplies the required instruments; the level of precision here is ±0.5 millimeters. Future developments at the EPFL University in Lausanne, Switzerland, will offer even more advanced sterilization, the integration of force sensors, and new areas of application of the robot.

Another stereotactic neurosurgical robot has been developed by IMMI medical robots in Grenoble, France. NeuroMate is an automatic and precise positioning system that enables pre-operation planning using simulation, and supplies constant feedback from the operation site.

Advance in New Dimensions of Surgery

In the Virchow Clinic in Berlin, Germany, professionals at the Clinic for Oral and Facial Surgery are concerned with the technical requirements for implementing robotic systems. They are researching robotic systems for possible application in hyperthermia (overheating therapy) and in the field of oral and facial surgery.

Many aspects need to be considered when integrating the robot, such as medical planning, technical planning, incorporating the robot, and the actual treatment itself. Various components are at the disposal of the Charité-Virchow Clinic for this: a mobile CT scanner, a planning and navigation system, and parallel as well as serial kinematics. The platform of the Motorized Surgical Scopes (MSS) with parallel kinematics, manufactured by Elekta, is affixed to the ceiling and therefore has an extensive working area. Also, a PUMA 560 robot is utilized for direct observation using the CT scanner. Experiments to date have been limited to phantom skulls; clinical studies on patients are being prepared.

Neurosurgical robot, screen shot, fixing framework, MINERVA, EPFL, Switzerland

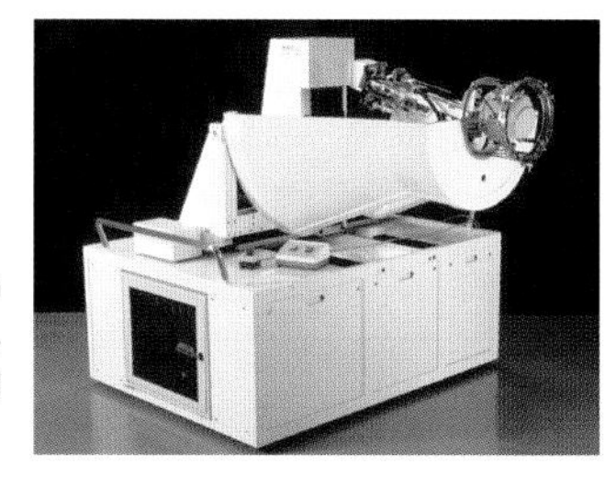

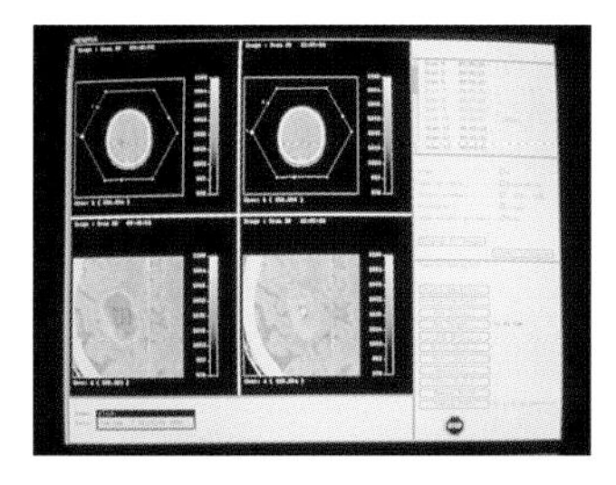

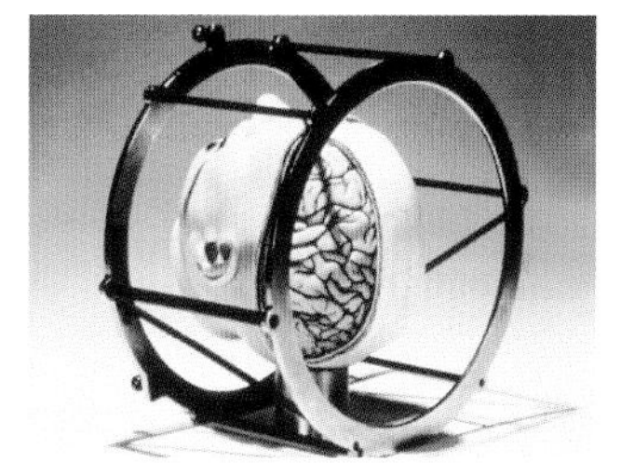

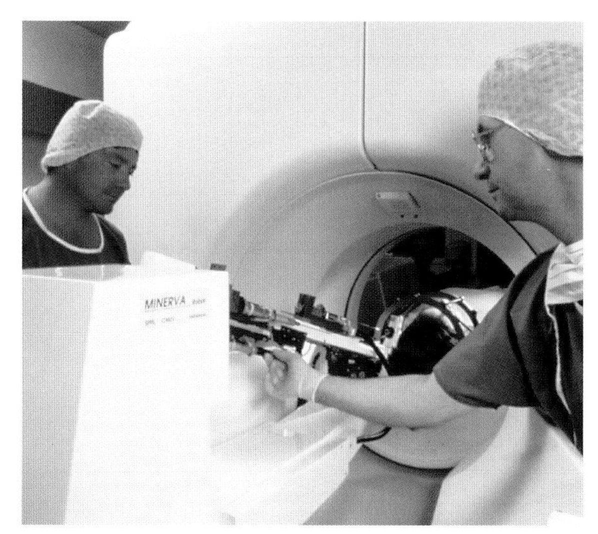

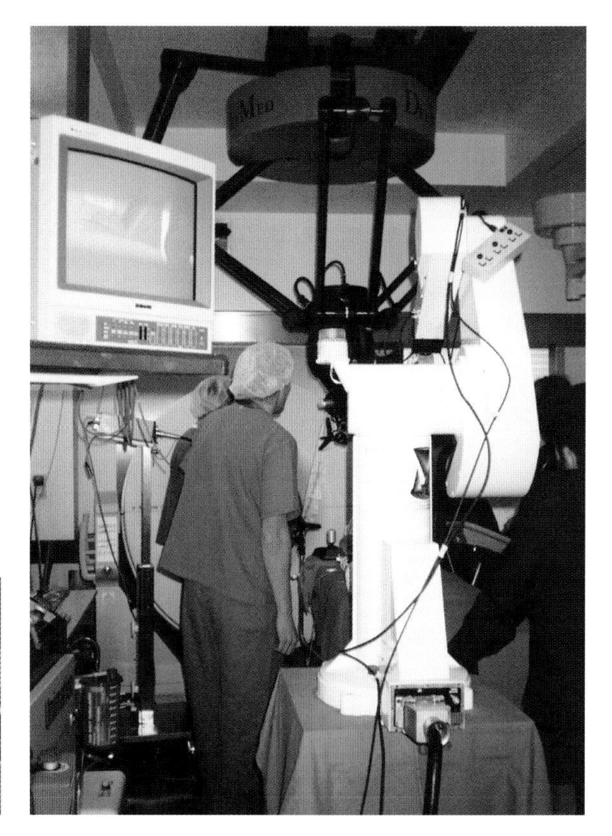

Virchow Clinic for Oral and Facial Surgery

Micrometer Precision Even Under Stress

The endoscope and navigation robot Operation System 2015, which has been developed at Fraunhofer IPA in Stuttgart, Germany, is an example of a guided intra-operative robotic system. There were two objectives for its development:

- To assist operations in the sub-millimeter range. Operations will become safer, and many new types of treatment will become possible.
- To investigate new possibilities for robot-assisted procedures in neurosurgery, ear, nose, and throat surgery, vascular surgery, orthopedics, and many other areas.

With the support of the branch of medical device technology at Siemens AG, Fraunhofer IPA, together with the Dr. Horst-Schmidt Clinics in Wiesbaden, Germany, and the reform design company in Stuttgart, Germany, has developed prototypes of an operation robot and an operation cockpit. These prototypes are system components belonging to a mechatronic assistance system. The system is universally applicable to medical devices and, due to its kinematics, can execute highly precise movements even when subjected to considerable stress.

A hexapod is used as an operation robot, the kinematic concept of which is based on the "Stewart Platform." With this type of construction, extremely high levels of precision can be achieved in the micrometer range, as well as a high degree of rigidity with a very compact set-up. An endoscope or other surgical instruments are affixed to the robot's instrument platform. The robot can be moved with the aid of a sled along a C-shaped carrier and can be swiveled around it. Using this method, the robot will be able to assist during most types of surgery.

Stewart platform
A parallel structure for a mechanism with six degrees of freedom which carries a work platform

Hexapod
Advantages of the hexapod system: All six degrees of freedom possessed by a rigid body in an open space can be reproduced. In comparison to conventional positioning systems with 6 rotatable joints, the construction of hexapods is highly compact and the area needed for installation is extremely small.
Other advantages of a hexapod's arrangement lie in the precise positioning ability of the tool: An accuracy of ±0.001 millimeters can be achieved. Hexapod systems also possess a considerably higher overall rigidity in comparison to conventional multi-axis systems and are therefore predestined for use in medical device technology where movements need to be executed in the sub-millimeter range.

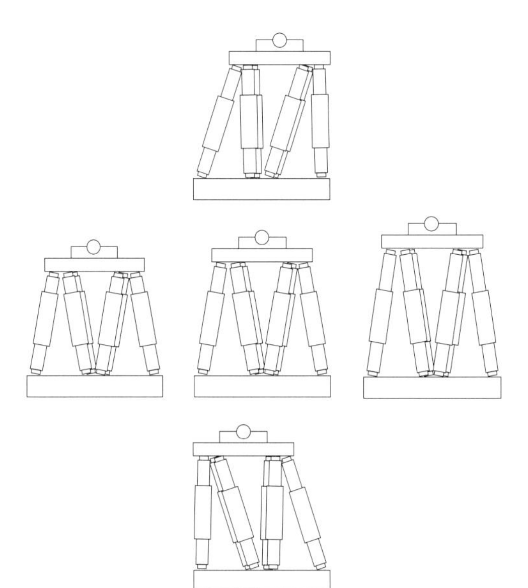

1996 Design study

1997 Development
Fraunhofer IPA/ partners

1997 Prototype

2000 Planned introduction onto the market

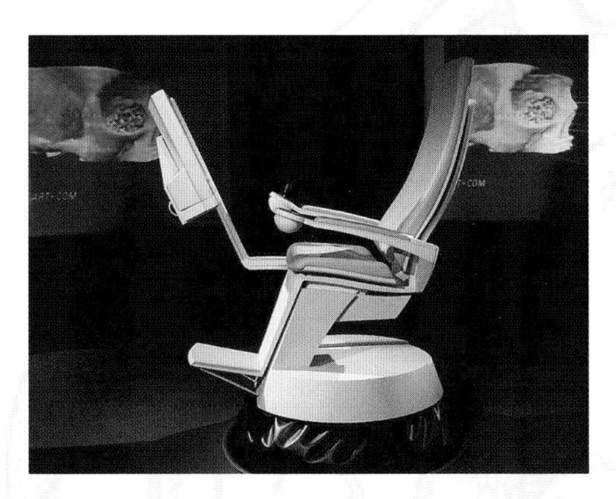

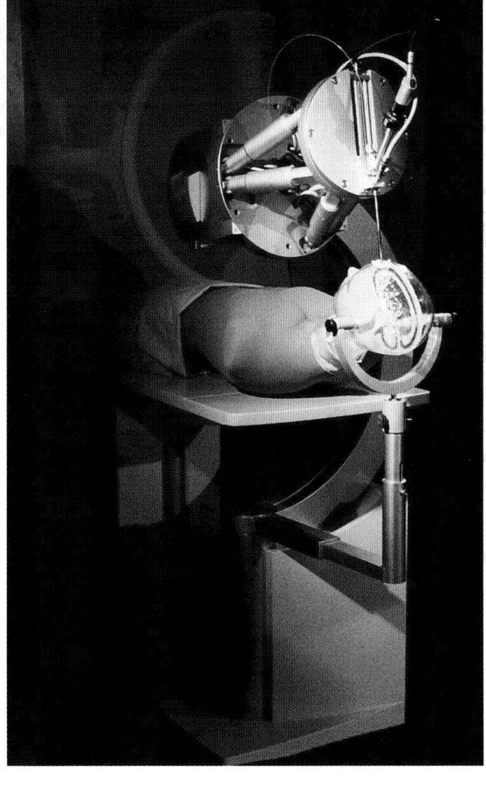

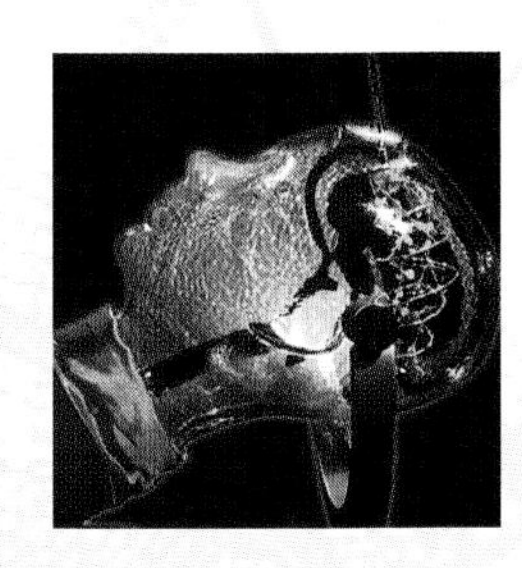

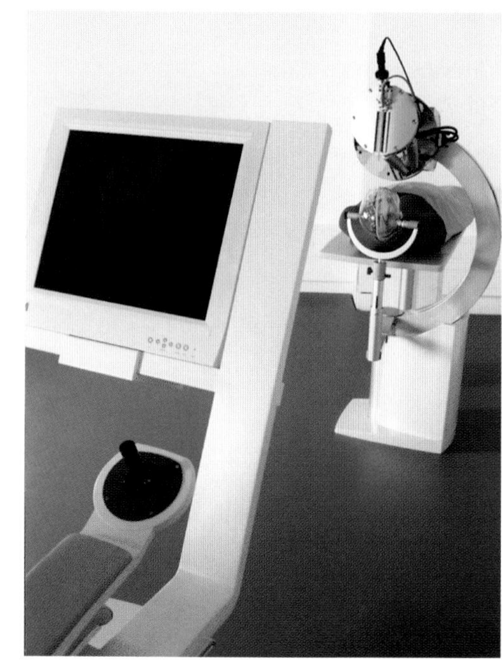

Operation cockpit with joystick, OP 2015

Operation cockpit, OP 2015

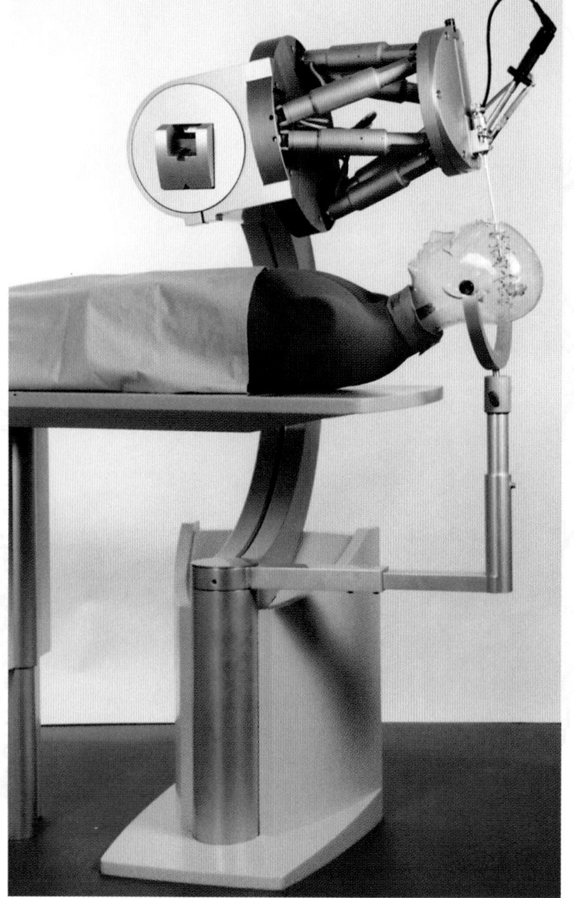

Hexapod robot, OP 2015

Kinesthetic Experiences in Sub-Millimeter Surgery

The surgeon controls the operation robot from an ergonomic operation cockpit similar to those in flight simulators. This helps him to operate the system utilizing haptic feedback. The operation cockpit is mounted on a hydraulic hexapod which allows movement to be represented in six degrees of freedom. The operation hexapod functions with an absolute accuracy of ±20 micrometers; a repeatability of ±2 micrometers can be achieved.

Dr. Urban of the Dr. Horst-Schmidt Clinics, Wiesbaden, Germany, had a vision of a "tangible flight" through the virtual anatomy of a patient. The visual and tactile feedback enables this vision to become a realistic experience. By scaling between the sub-millimeter range of the operative world and the surgeon's macro-world, many surgeons expect to attain a more reliable perception of the actual endoscopic field. They obtain a better feel for the boundaries of the work space, which is 100 by 100 by 50 millimeters, and the speed of movements in critical areas.

The Surgeon Always Maintains Complete Control

The arrangement of the man-machine interface in the cockpit is particularly notable. The user interface has been designed to be easy and safe for a surgeon to use. The most important visual navigation aids, such as the image from the endoscope, have been integrated into the interface.

By integrating the control device into the armrest, its operation has become comprehensible and user-friendly. Currently, the robot and the endoscope are still controlled via a conventional industrial joystick. Ergonomic input devices modeled on the surgeon's arm and hand movements during an operation are currently being developed.

The target of the endoscope is pre-selected with the aid of a target symbol. For safety reasons, the operation robot's actual movements are only initiated when a key has been pressed a second time. Because of this, the robot cannot make unintentional movements.

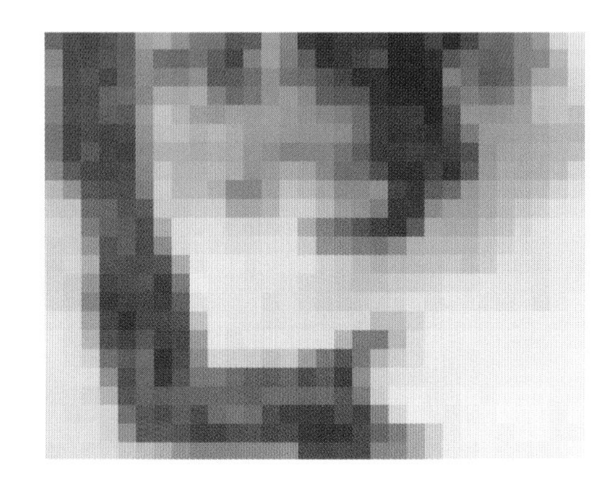

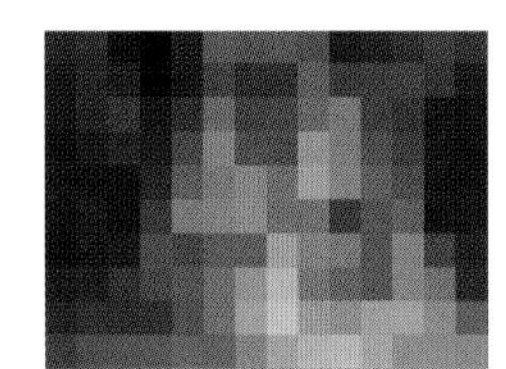
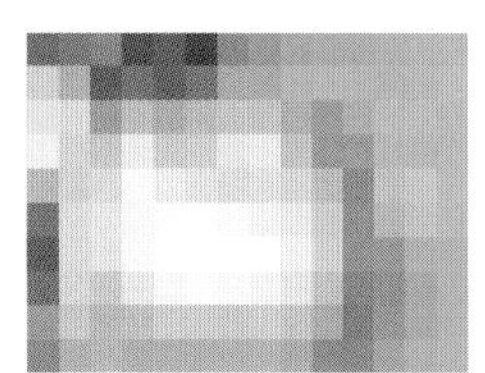
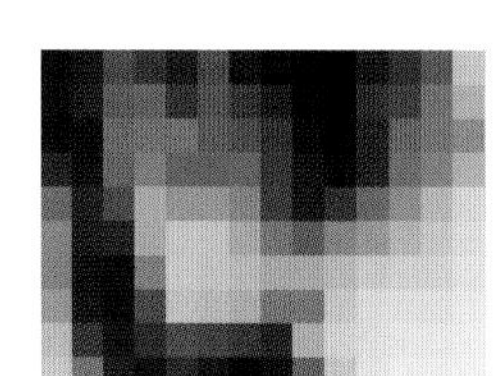
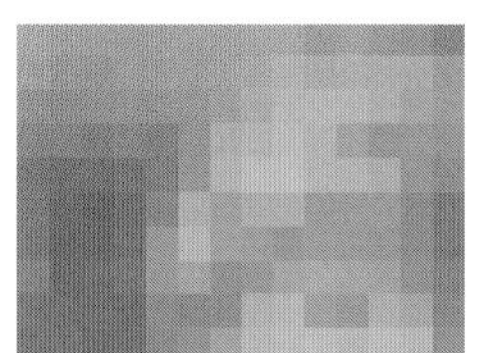

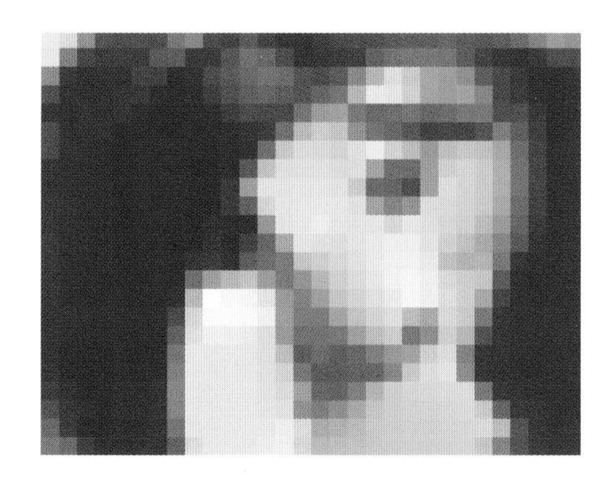

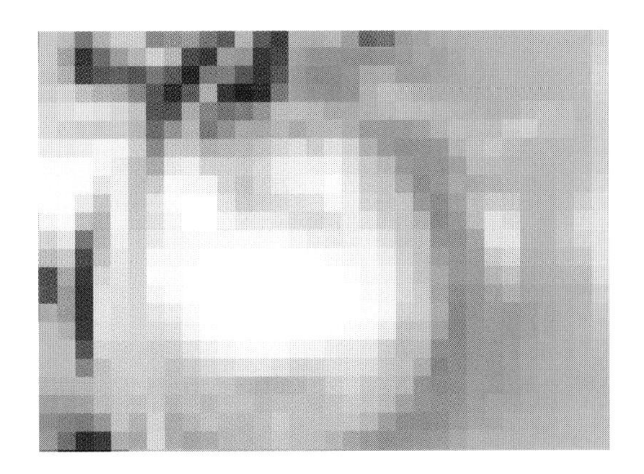

Underwater

Strong Forces in the Depths of the Oceans

The world's oceans cover the majority of the Earth's surface, yet these are the least well-researched parts of our planet. There are creatures living several kilometers deep which man has never seen face-to-face. Raw materials are suspected to be here, the mining of which could become increasingly important within the next few decades.

Humans Are Not Suited to Underwater Conditions

A person can only stay below the surface of the water for a short period of time, even if he or she has an oxygen supply. Even with a diving suit, only a certain depth can be attained because of an increase in hydrostatic pressure as depth increases. Without technical assistance, the ocean floor will no doubt remain concealed from our eyes.

Even if technical aids enable a diver to stay beneath the water for some length of time, the diver would not be able to work very easily there because oxygen tanks and diving suits are a severe handicap.

Underwater service robots are therefore ideal helpers: they undertake the jobs that would be too dangerous or simply impossible for humans. Included here is offshore work, such as on drilling platforms or underwater pipelines; inspection work in docks; and work in other hazardous underwater areas, such as in nuclear power stations or pipelines.

An underwater robot is a small submarine with an onboard computer, camera systems, searchlights, and sophisticated sensors. These robots sometimes have work systems, such as manipulator arms. Supply lines link the unmanned underwater vehicle to a control station in a sheltered area. In this way, these teleoperated robots can perform a whole range of tasks, such as measuring, inspecting, monitoring, and handling objects.

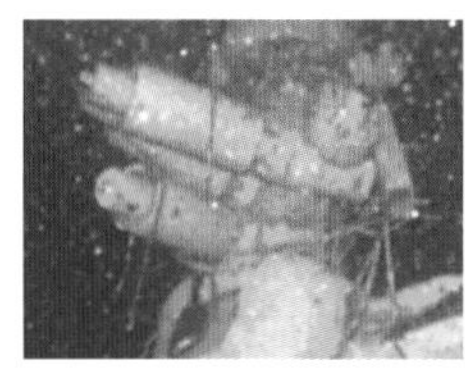
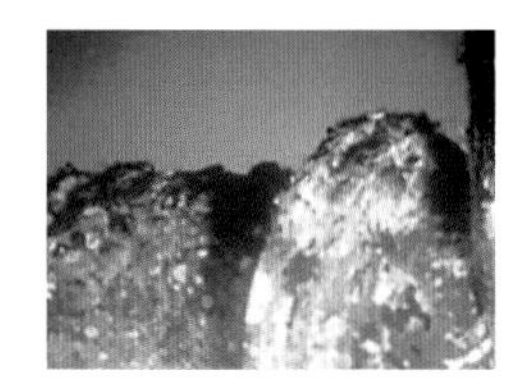

Potential operational areas for underwater robots – offshore, docks, underwater research

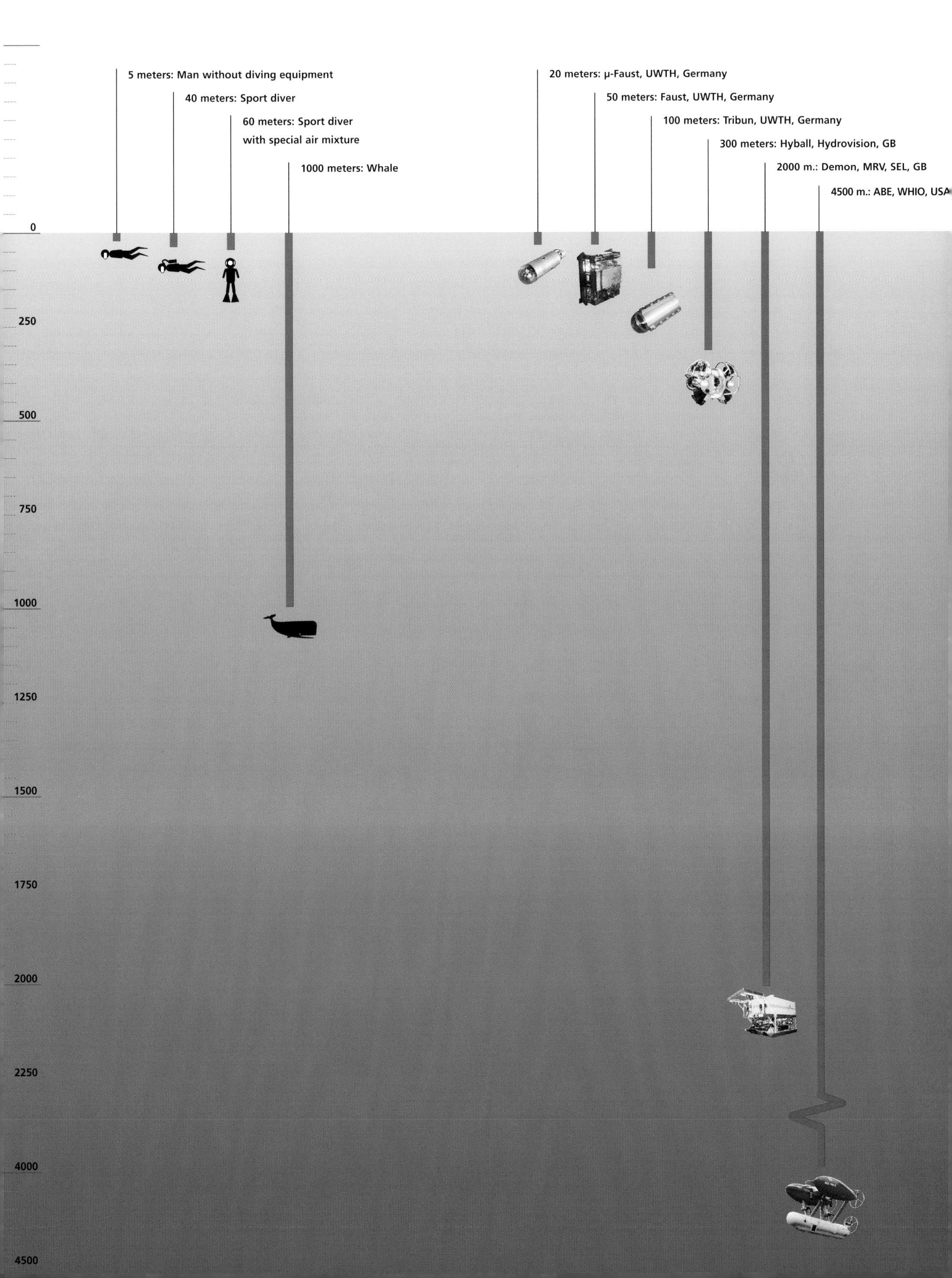
5 meters: Man without diving equipment
40 meters: Sport diver
60 meters: Sport diver
with special air mixture
1000 meters: Whale
20 meters: µ-Faust, UWTH, Germany
50 meters: Faust, UWTH, Germany
100 meters: Tribun, UWTH, Germany
300 meters: Hyball, Hydrovision, GB
2000 m.: Demon, MRV, SEL, GB
4500 m.: ABE, WHIO, USA
0
250
500
750
1000
1250
1500
1750
2000
2250
4000
4500

A Job for Faust!

teleoperable
Remote-controlled

trim
Adjustment of a boat's floating position in a longitudinal direction either by redistributing the load or the fixed ballast (sand, iron bars), or by filling or pumping out the trim tank.

Faust (Free-diving Autonomous Underwater System carrier) was developed at the UWTH, a department of the Institute for Material Science at Hannover University, Germany. The teleoperable underwater handling system is utilized in underwater installations of nuclear facilities to carry out inspections, assembly, and dismantling tasks.

Faust has a maximum diving depth of 50 meters. The service robot can be controlled in five axes. By using its eight electrically operated propellers and its trim, which is independent of its pose, it can work at any angle in any chosen direction. Its caterpillar drive enables it to approach almost any surface accurately.

An encapsulated onboard computer performs the steering tasks and an external computer acts as the operator interface. The operator controls the robot's movements by using a keyboard and a multifunction track ball. He or she only has to stipulate the underwater robot's desired pose; the onboard computer then calculates the control commands for the drives.

The supply and signal lines are all bound together in one cable, which also functions as a salvaging line if the system should break down. Thanks to its compact aluminum construction (790 by 700 by 305 millimeters), the vehicle only weighs 42 kilograms.

Cameras, searchlights, and sensors enable Faust to perform complex inspection and measurement tasks beneath the water. The operator can monitor the work and movements by way of a camera image. A tool platform with two prismatic joints is integrated into the carrier system, and many tools can be attached to it, including various sensors, a small plasma cutting torch for carving up steel components, or a gripper capable of carrying a maximum payload of 1 kilogram.

Underwater cutting, FAUST, UWTH, Germany

Small µ-Faust for Precision Work

µ-Faust has been developed for carrying out work in or on components with small cross-sections. This cylindrical submarine, which is 50 centimeters long and 12 centimeters in diameter, carries a swiveling CCD camera inside its dome-shaped Plexiglas head. µ-Faust is of modular construction, and by installing additional segments, it can be adapted to fit completely different requirement profiles.

A propeller advances the robot, and transverse rudders permit spatial orientation alignment. It has a maximum diving depth of 20 meters. As is the case with Faust, communication between the operator and the onboard computer takes place by way of an external computer. The submarine's movements are given by the operator who can choose between using a joystick or using a mouse; the onboard computer then calculates the corresponding changes for the drive.

The underwater robot Tribun (Tribun stands for "partially automated robotic system for underwater inspection and tasks"), also originated in Hannover, Germany. It represents a further development of the µ-Faust, but with this model more emphasis has been placed on maximizing modularity and user-friendliness.

The four plug strips placed around Tribun accept sensors and actors, which can then be connected to the onboard computer via a CAN bus. The submarine is capable of diving up to 100 meters below sea level.

CCD
Abbreviation for "Charged Coupled Device". CCDs transform light into an electrical charge; for this reason, they are used in the form of chips in scanners, digital cameras or camcorders.

CAN bus
Controller Area Networks
Highly reliable digital data transfer with a high data rate between sensors and actors (on field bus level)

Tribun, UWTH, Germany

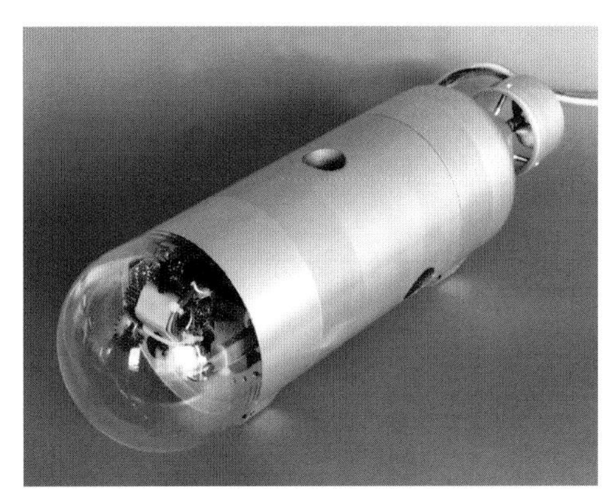

µ-Faust , UWTH, Germany

Inspecting the Hulls of Ships

The Scottish company Hydrovision, based in Aberdeen, manufactures Remotely Operated Vehicles (ROV) for underwater applications. These underwater robots carry out inspection work, such as checking ships' hulls, harbor walls, and waste-water supplies, as well as surveillance jobs, such as monitoring divers or underwater drilling sites. ROVs are also used for carrying out salvaging tasks or minor repairs.

translational
Progressive movement in a straight line

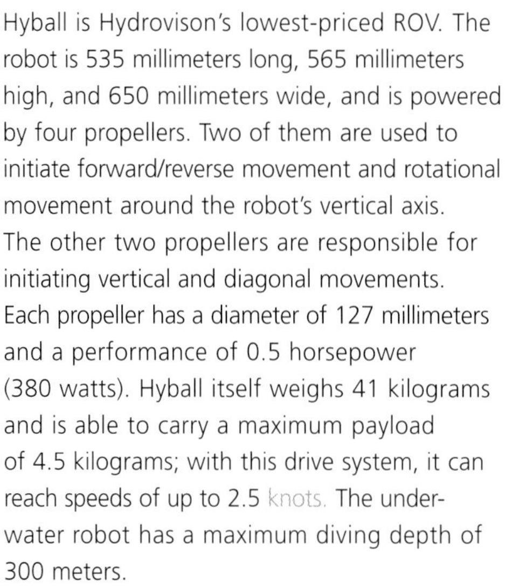

Hyball is Hydrovison's lowest-priced ROV. The robot is 535 millimeters long, 565 millimeters high, and 650 millimeters wide, and is powered by four propellers. Two of them are used to initiate forward/reverse movement and rotational movement around the robot's vertical axis. The other two propellers are responsible for initiating vertical and diagonal movements. Each propeller has a diameter of 127 millimeters and a performance of 0.5 horsepower (380 watts). Hyball itself weighs 41 kilograms and is able to carry a maximum payload of 4.5 kilograms; with this drive system, it can reach speeds of up to 2.5 knots. The underwater robot has a maximum diving depth of 300 meters.

knots
Abb. kn., originating from the markings on a log line to denote the speed of a ship; 1 kn. = 1 sea mile/hour = 1,852 km/h

Hyball has a rotatable chassis and the platform of the standard model comes equipped with a wide-angle lens CCD camera which can be swiveled a full 360°. The camera support can be equipped with up to three cameras. In addition, Hyball possesses two fixed, forward-facing 100-watt quartz-halogen searchlights and a further two 75-watt searchlights on the camera support. The standard sensor equipment consists of a depth gauge, a magnetic compass, a gyroscope for directional determination, and a leakage detector. The robot is controlled by a 3-DOF (degrees of freedom) joystick. Hyball can be optionally equipped with a simple manipulator arm affixed with a three-finger gripper and one translational degree of freedom.

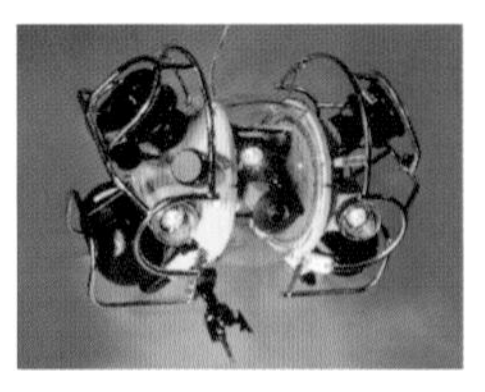

The Offshore Hyball varies from the standard version in that it is capable of carrying a heavier payload and offers a broader range of additional functions. It is not intended to replace Hyball; rather, it should perform more demanding work conditions. Offshore Hyball is intended to be used in the offshore oil and gas industries.

Offshore Hyball
- **propeller**
- **gripping arm**
- **camera unit**

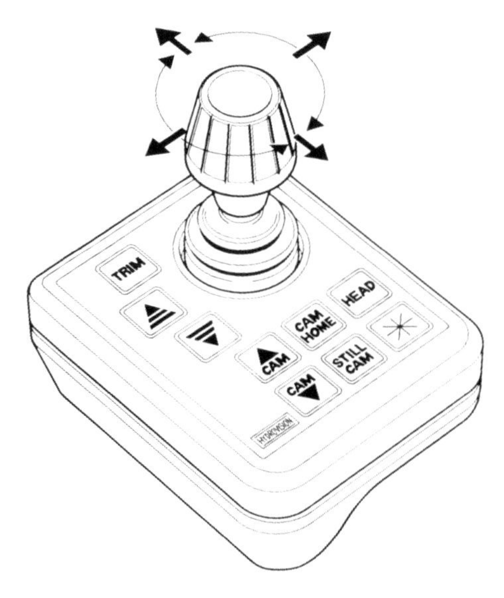

3-DOF joystick for controlling the robot

Diablo, Hydrovision, GB

Demon

Control center

Diablo and Demon Perform the Hardest Work

Diablo and Demon are the names of the two most powerful ROVs in Hydrovision's range of products. They both work in depths of up to 2,000 meters. Demon carries a payload of 300 kilograms and weighs 2.8 metric tons. Diablo can be adapted to carry up to 450 kilograms; this underwater robot then weighs over three metric tons.

Diablo and Demon have been designed for use in rough conditions and are ideally suited for assembly and monitoring purposes. This type of high-performance robot is used on drilling platforms and in underwater pipeline construction work. The robots can be equipped with manipulator arms and up to five different camera systems.

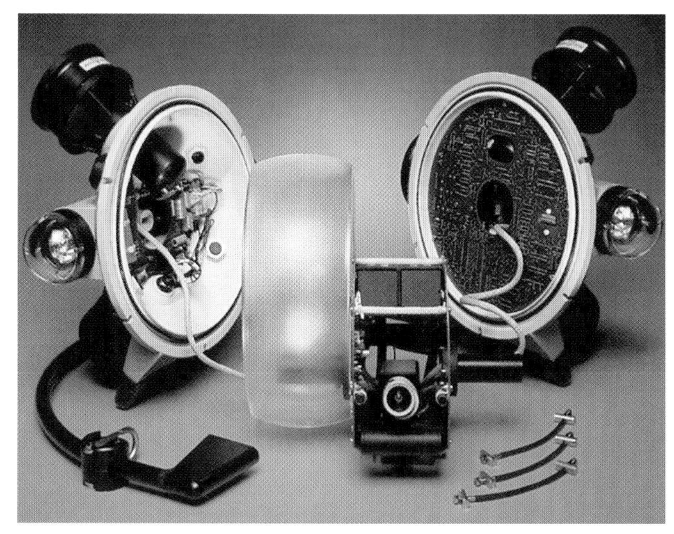

Hyball's inner workings, Hydrovision, GB

Diablo's drive system, Hydrovision, GB

Robot Weighs Five Tons

MRV 4, SEL, GB

Slingsby Engineering Ltd. (SEL), based in Kirkbymoorside, England, is one of the leading manufacturers of ROVs and underwater manipulators. The modularly constructed underwater robot MRV (Multi-Role Vehicle), is at the top of SEL's production range. Work packages to suit a wide variety of requirements can be integrated into the basic version, provided a maximum system weight of 5,000 kilograms is not exceeded. For use as work or measuring systems, two manipulator arms, five camera systems, and a sonar device, among other options, can be attached to the carrier system. The 3.5 kilowatt-strong searchlights ensure excellent illumination of the working area. The standard ROV, which can be up to several meters long depending on how it is equipped, can reach depths of 1,000 meters, and an option is available to even extend this to 2,000 meters.

SEL also develops and manufactures manipulator arms specially adapted to deep-sea diving requirements, which can be attached to the ROVs.

These multifunctional robotic arms can activate valves, operate hydraulic tools, handle sensors, and carry out maintenance work under water. The variations in kinematics allow between 5 and 7 degrees of freedom and maximum extended lengths of the arms to over 2 meters. The joints are controlled using hydraulic cylinders. The arm segments are made of corrosion-resistant, high-tensile aluminum alloys, and the connecting elements are made of high-grade steel. These robots can handle loads of up to 200 kilograms.

Two different control variations are available: position feedback (control of the parameter of position) and rate control (control of the parameters of position and velocity).

Diving procedure

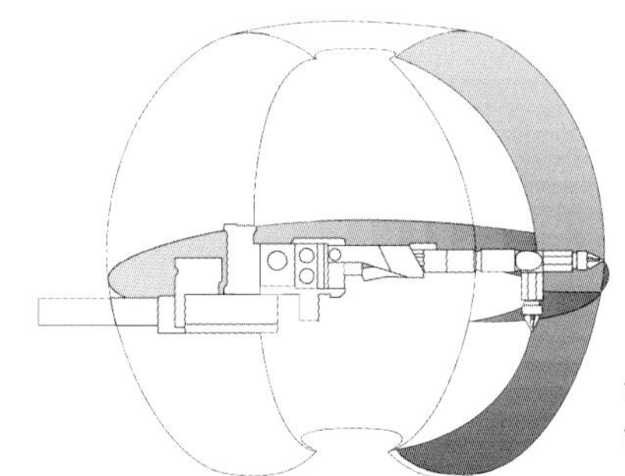

Work space of the manipulator

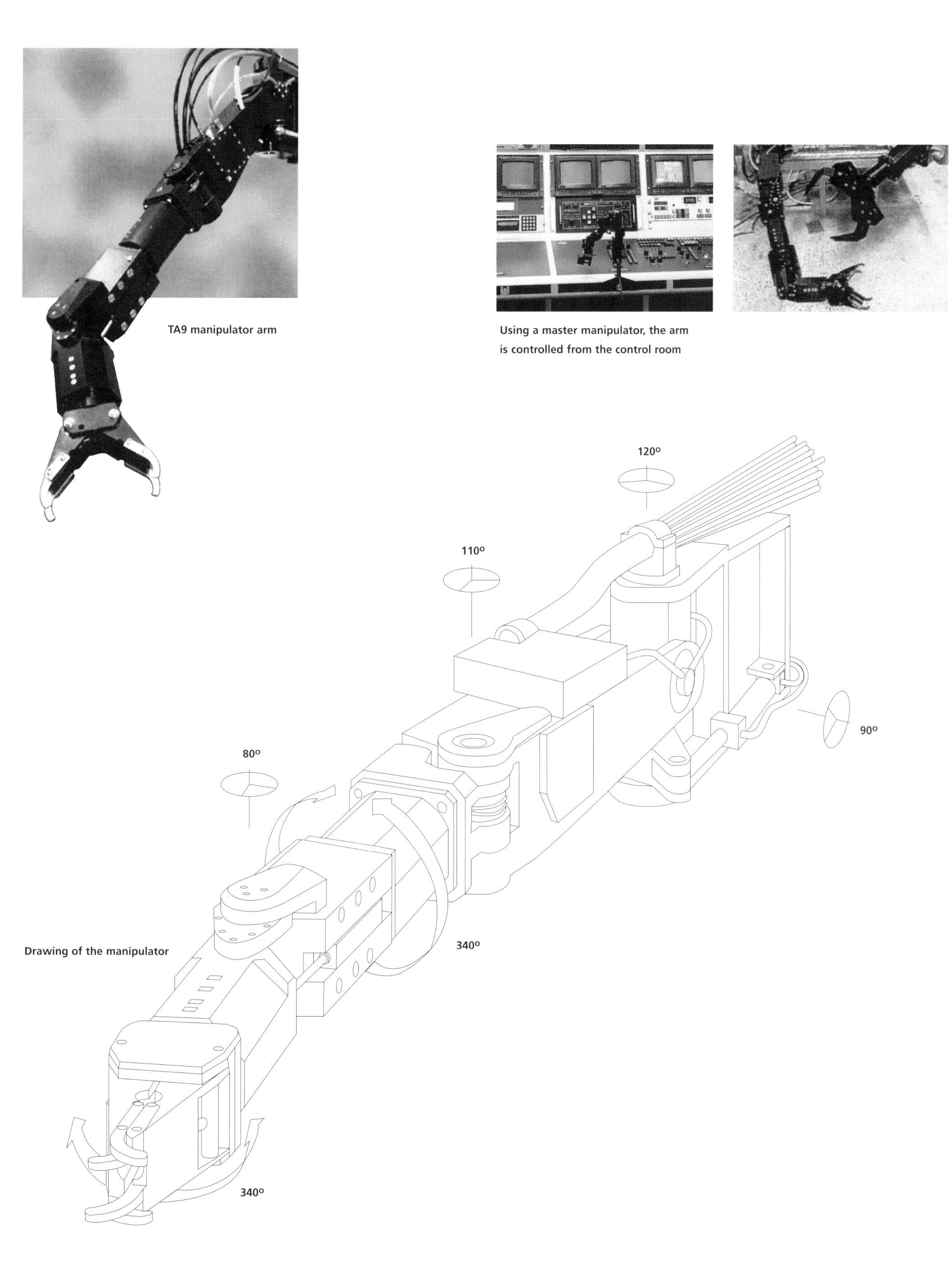

TA9 manipulator arm

Using a master manipulator, the arm is controlled from the control room

Drawing of the manipulator

Deep-sea Research Using Ultrasound Navigation

One of the few autonomous underwater robots not used for military purposes is called ABE (Autonomous Benthic Explorer). ABE was constructed at the Advanced Engineering Laboratory of the Department of Applied Ocean Physics and Engineering at the Woods Hole Oceanographic Institution (WHOI). The college, based in Massachusetts, is renowned for its unique developments in the field of ocean research and underwater robotics.

ABE can dive to a maximum depth of 4,500 meters to carry out long-term experiments. Depending on the batteries with which it has been equipped, the autonomous diver has a range of between 10 and 100 kilometers and is capable of reaching speeds of up to 2 knots. The robot navigates with the aid of two or more transponders in a network of acoustic signals and follows the contours of the ocean floor, floating just a few meters above it. Using a minimum of energy, ABE glides to a prescribed position far below the surface.

At present, ABE is equipped with a stereo single-image video camera, as well as sensors for determining salt concentrations, temperature, magnetic fields, and distances to the ocean floor. Additional measuring devices can easily be integrated. So far, ABE has been implemented for measuring magnetic anomalies occurring near the ocean floor at the transitions of tectonic plates.

ABE's design engineers are planning to go a step further: at the moment they are performing research on the next generation of AUVs (Autonomous Underwater Vehicles), which will be capable of working for almost unlimited periods of time, able to supply themselves autonomously with energy from underwater docking stations.

ABE, WHIO, USA

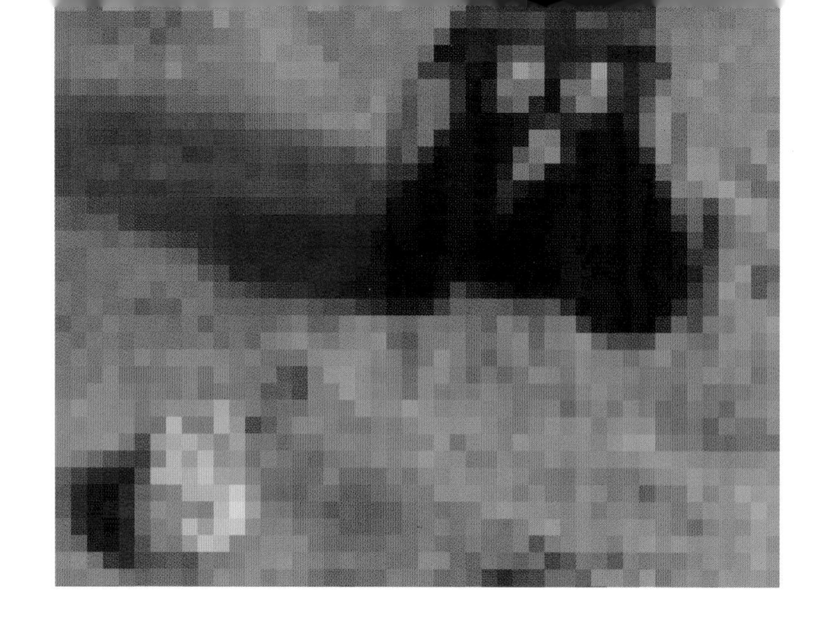

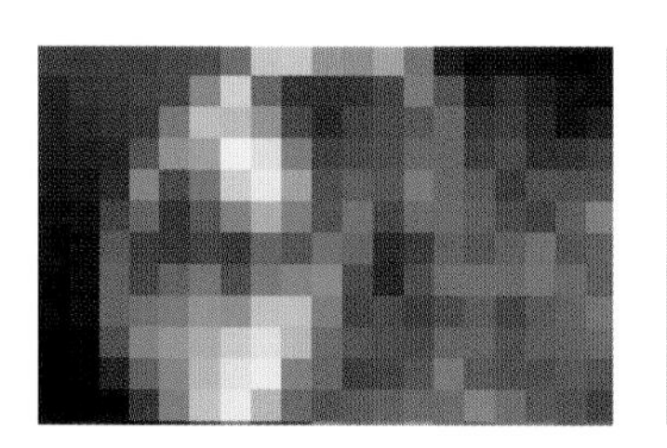

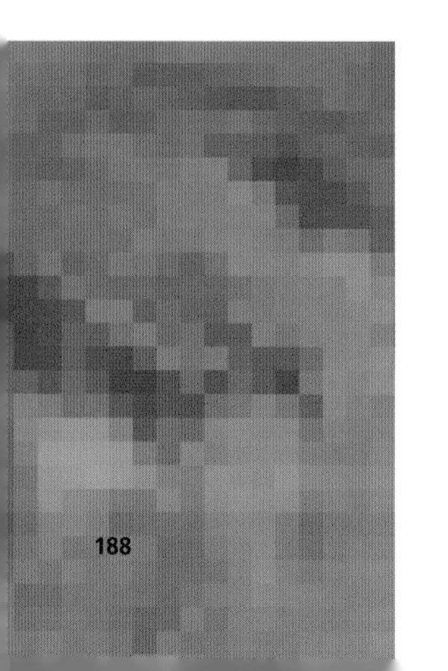

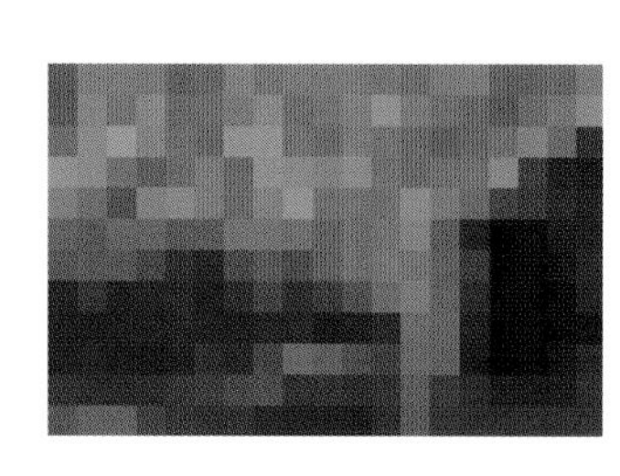

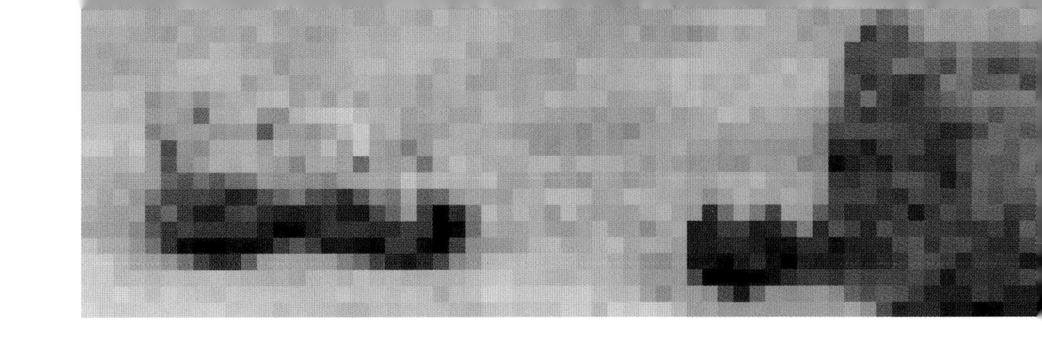

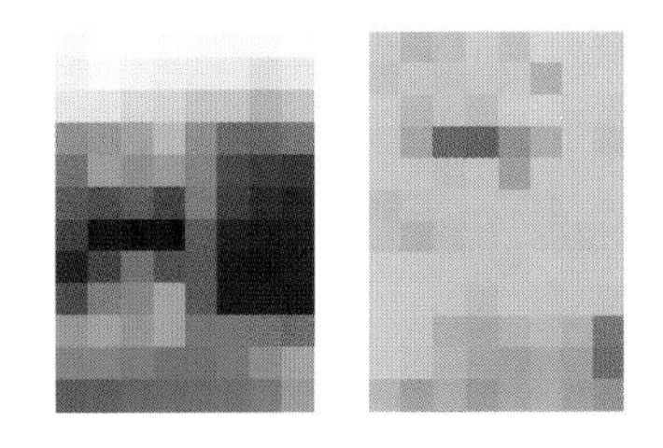

Space

Odyssey 2000: Completely Autonomous

HAL and R2D2 had quite a presence in science fiction films. But robots have a firm place even in real space travel. It would be impossible to explore distant planets, maintain permanently available orbiting stations, or conduct complex and lengthy experiments in space without intelligent, remote-controlled autonomous vehicles and automatic machines.

The functional requirements of manipulators and robots in space are wide-ranging. They demand a hierarchically structured and modular automation concept that can be adjusted to suit the particular application. Humans must be taken into account as far as the various levels of control, monitoring and decision-making are concerned: everything between telemanipulation alone and autonomous robotic operation should be possible.

Lightweights with a Gentle Touch

Lightweight robotic systems are the goal of space travel robotics. On the one hand, they should be an extension of the human arm, that is, they should be remote-controlled and remotely programmable by payload specialists. On the other hand, by using many different sensors, they should be able to function with maximum autonomy so that they can carry out sub-tasks independently.

By further developing complex, two-jaw grippers to become finely jointed, multifingered hands, a robot's manipulative skills become closer to a human's dexterity. As it is, humans are severely restricted when carrying out external tasks in open space. But service robots don't just perform important tasks in space, they also carry out work on Earth essential to the safety of manned space travel.

Space shuttle Endeavor lifting off, NASA

Tests of Vital Importance on Earth

Tessellator, tile inspection robot, CMU, USA

The space shuttle's heat shield is composed of 17,000 small tiles. They shield the shuttle's underside and nose from heat, and prevent the shuttle from burning up like a meteorite on its return to Earth. This heat shield must be painstakingly checked over before each takeoff to ensure that everything goes smoothly on reentering the atmosphere.

For the first few space shuttle missions, this check was carried out manually. From the moment the space shuttle entered the Kennedy Space Center right up until take-off, the heat tiles had to continuously be checked over visually and had to be chemically sealed to prevent moisture absorption. The workers had to wear heavy protective clothing and face masks because of the toxic chemicals that needed to be applied.

Because all the other inspection and maintenance work also had to be carried out on the shuttle simultaneously, there was quite a crush in the immediate vicinity of the orbital glider; cables and hoses were always blocking the way. In some places, the clearance space beneath the shuttle is as little as 1.75 meters. Due to the limited freedom of movement caused by the protective clothing and the cramped space, this was particularly strenuous work for the personnel.

To automate this task, a mobile robot has been developed in a NASA project at the Carnegie Mellon University (CMU) in Pittsburgh. The Tessellator consists of a mobile platform with an integrated manipulator arm. The mobile platform is of a particularly rigid construction, giving the robot the stability required for the high level of accuracy demanded of it.

Image-Assisted Task Planning

An onboard computer controls Tessellator's process tasks, whereas the arm and wheel movements are regulated by a low-level control unit with amplifiers. Two additional computers control both the camera system and the injection system for the sealing fluid.

Before each inspection, Tessellator is informed of the shuttle's position and of the inspection sequence to be executed. Based on this data, the mobile robot scans its environment and determines its actual position. Its camera locates the tile to be checked. In the process, Tessellator divides its work area into small, equally-sized segments in order to minimize overlapping.

Tessellator adjusts the movement of its manipulator arm to the shape of the orbital glider. To ensure the required stability of the system, the height of the robotic arm is pre-adjusted by a platform actuated by a prismatic joint. Fine positioning takes place by way of a small linear axis. The manipulator arm carries either a camera for visual checking, or the sealing tool.

By comparing the tile being examined with a neighboring tile, Tessellator is able to recognize anomalies such as cracks, scratches, or discoloration. If it is uncertain whether the tile it has checked is undamaged, the robot stops, and a worker can then inspect the tile again on the screen of his monitoring computer. At the end of an inspection sequence, Tessellator sends its measurement data to a NASA central database.

The next generation of space shuttles will cost Tessellator its job, because the new shuttles won't need heat-resistant tiles. Even service robots are made redundant.

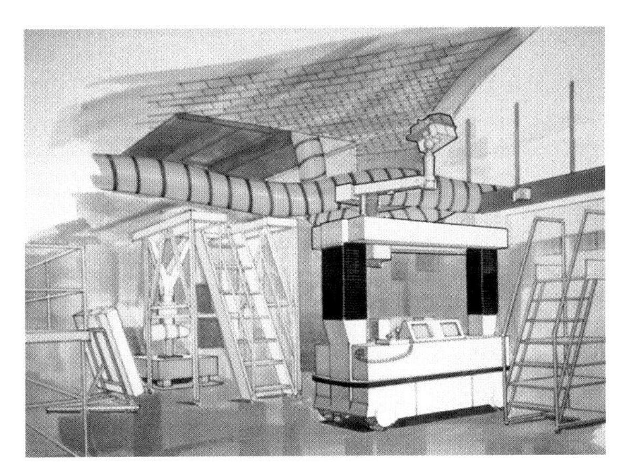

Design sketch of Tessellator, CMU, USA

Moon Walker

Sponsored by NASA, the Lunar Rover initiative is another of the CMU Robotics Institute's main development themes. Its aim is to land a research robot on the moon before the end of the decade. Sandia Laboratories, LunaCorp, and other governmental and industrial facilities are taking part in the project, as well as NASA and the CMU.

Dante was the first sub-project carried out by this community enterprise – a prototype for exploring inhospitable planets. In January 1993, Dante I attempted to explore the crater floor of Mt. Erebus in Alaska. Although the eight-legged walking robot had to give up just before achieving its goal, it still obtained important information that was useful when developing the successor device, Dante II.

This second robot was given bigger legs, a different leg and gait configuration, and a more stable cable as a safety line. Instead of having four legs on each side, Dante II was built with four front legs and four back legs. This configuration gave the robot even more stability and simplified the control algorithm. The improved legs were able to carry three times as much weight as Dante I's.

In July of 1994, Dante II proved its ability as it descended into the crater of the still-active volcano Mt. Spurr, in Alaska. This time, it was possible to test the system's highly developed telecommunications and control software. The walking robot's sensors supplied measurement data on the content of carbon dioxide, hydrogen sulfide and sulfur dioxide in the volcanic gases.

Research robot Ambler, CMU, USA

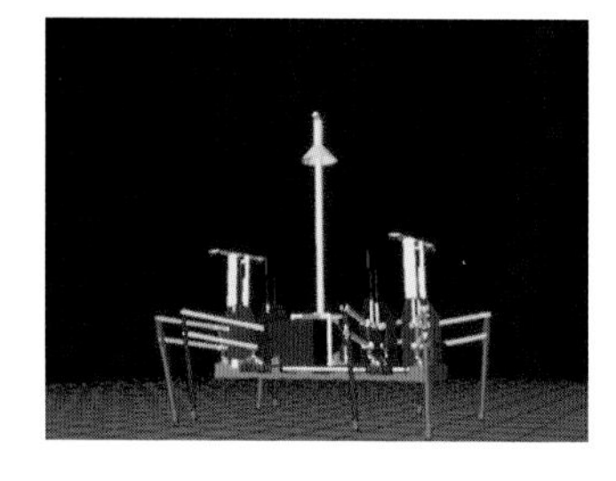

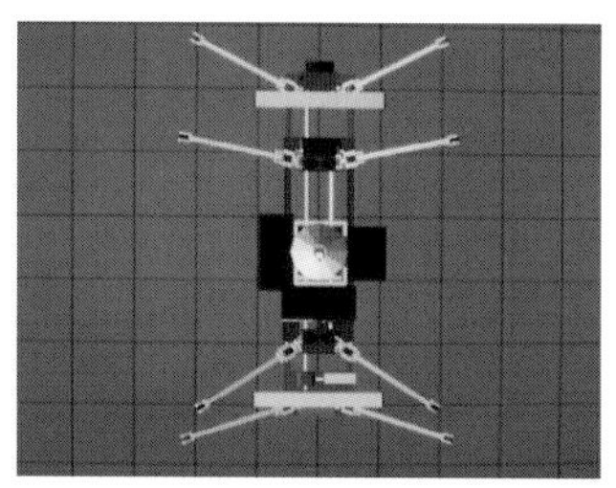

Computer simulation, CMU, USA

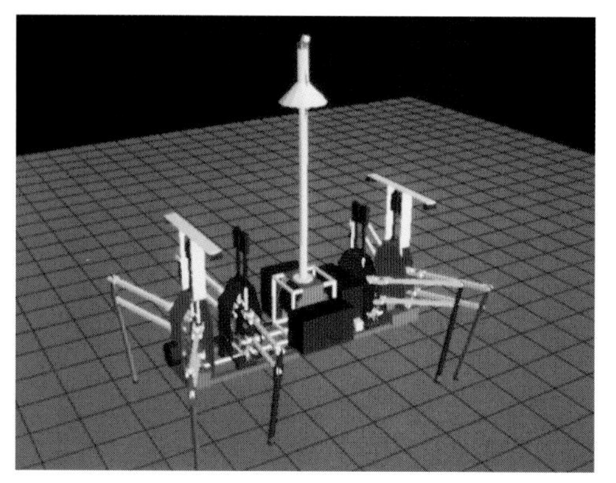

Research robot Ambler, CMU, USA

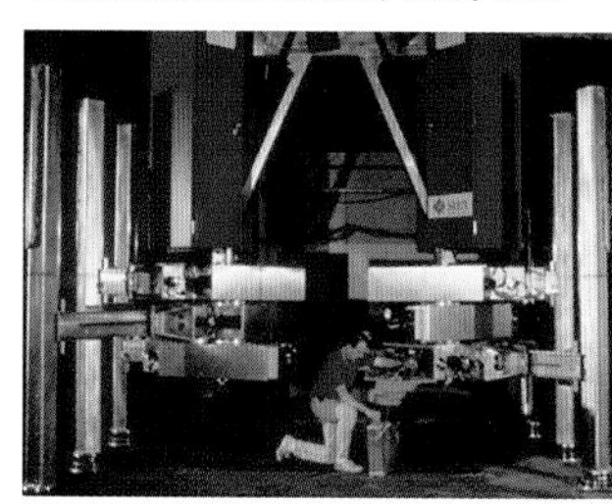

Dante II being transported to its operational site in Alaska

Research robot Dante II, CMU, USA

Planet Rover Practices in the Desert

Design sketch of a planet rover, CMU, USA

Another project sponsored by the NASA was the Atacama Desert Trek, completed in 1997.
In this project, it was intended that Nomad, the planet rover developed by the CMU, should travel a distance of 200 kilometers across the Chilean Atacama desert. In addition to exploring the landscape, which is similar to the surface of Mars or of the Moon, the testing of new communications and control technologies was also planned.

Nomad was supposed to travel over unknown terrain on its way across the desert, either autonomously or teleoperated from a ground station several thousand kilometers away. In order to cope with any terrain, the engineering designers gave the rover a chassis that could alter its tracks and wheel base to suit environmental conditions. Nomad, weighing 550 kilograms, also had 4-wheel drive and could turn on a dime thanks to its four steerable wheels.

Planet rover, Nomad, CMU, USA

Camera with All-around Vision

Because conventional video cameras have a limited field of vision, Nomad was equipped with an innovative "panospheric camera", which supplied high-quality pictures with an extremely wide image-angle. This camera was mounted vertically onto the roof of the vehicle and pointed upwards. A polished steel ball was suspended exactly on the image axis. The image recorded by the camera depicted the spherically distorted reflection of the rover's environment.

By using specially developed software modules, the distorted image could be corrected to obtain a panoramic image of the surrounding area obtained. A second conventional camera affixed toward the rear of the rover covered the area which, due to the position where the panospheric camera was mounted, could not be seen. As the rear section of the rover was not equipped with laser scanners, this camera also had an orienting function when the vehicle reversed.

The exact positional determination was carried out on the ground using a DGPS system (Differential Global Positioning System). Under ideal conditions, Nomad was able to determine its position on the ground to within ±20 centimeters with this equipment.

Planet rover, Nomad, CMU, USA

Intelligent Operation and Navigation

When exploring distant planets with telemanipulated robots, the rover's distance from the ground station plays an important role. Depending on the distance, several seconds can pass between the time that the signal is transmitted and when it is received by the robot, even though the signal moves at the speed of light. The decisive factor here is not only distance but also the number of relay stations involved, because these stations first analyze and check the signal before transmitting it further.

But this delay doesn't upset Nomad. Thanks to its sophisticated sensors, the robot is able to realize if it is on a dangerous route. Obstacles are recognized and plotted on a digital map. If the operator steers Nomad towards an obstacle by remote control using the "safeguarded teleoperation" mode, the rover ignores the control signal and uses its sensors to adjust its path. Once the rover has completed its avoidance maneuver, the ground station regains steering control.

Nomad's success depends on its individual modules being able to work together optimally, for example, the onboard computer, camera system, or communications system. Therefore, the autonomous robot constantly monitors the "vital functions" of its components and prevents the operator from giving incorrect directions when in teleoperated mode. For this reason, Nomad limits its maximum speed to 0.2 meters per second.

In June and July of 1997, Nomad completed its first mission, which was in Chile. The rover's next trips took it to Antarctica.

Marslander MVACS, JPL, USA

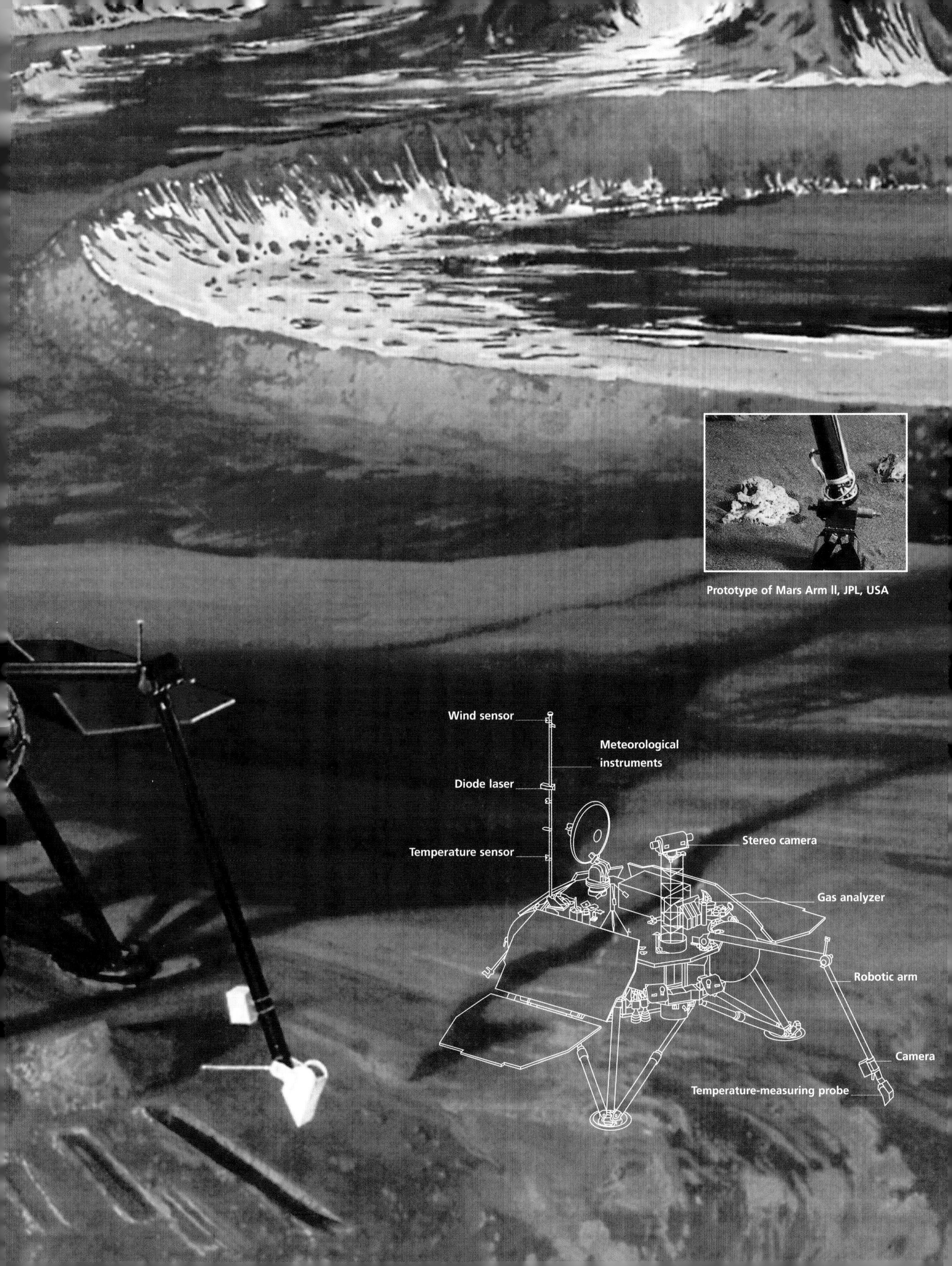

Prototype of Mars Arm II, JPL, USA

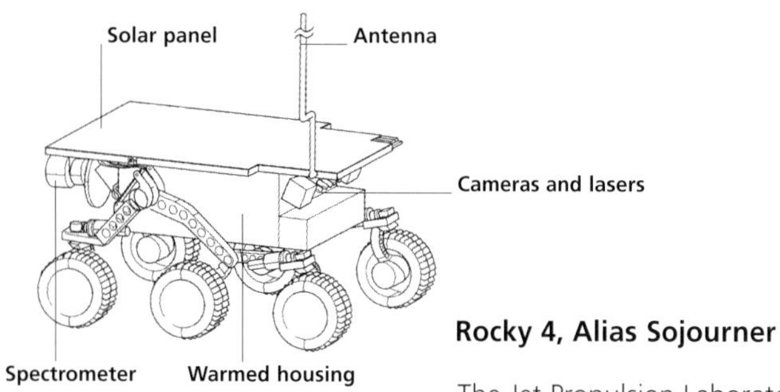

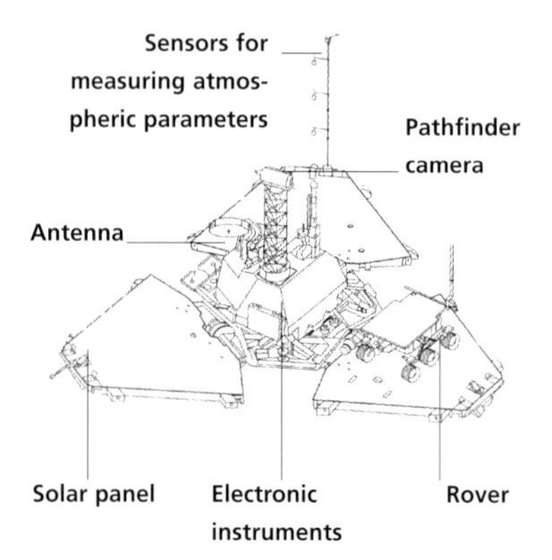

Rocky 4, Alias Sojourner

The Jet Propulsion Laboratory (JPL) is a research center sponsored by NASA and belonging to the California Institute of Technology in Pasadena. It is involved in all the US authorities' space missions. The Mars rover Sojourner – known internally as Rocky 4 – was developed at the JPL, as was the Pathfinder mission's Mars landing craft to a large extent.

On the fourth of July in 1997, coinciding with Independence Day, the small remote-controlled vehicle had the world on the edge of its seat. On entering Mars's atmosphere, parachutes slowed the landing craft's speed sufficiently so that the airbags, which opened just before landing, were able to cushion the impact. Due to the low gravitational force on Mars (just 40% of the Earth's gravitation), the padded landing ferry bounced several times before it finally came to a standstill.

Once the air from the airbags had been expelled, the landing craft's ramp was opened by remote control, and Sojourner was ready to perform its mission. The main intention was that the six-wheeled vehicle should carry out geological measurements of the planet surface. The small vehicle's specially designed kinematics enabled it to turn on the spot and avoid obstacles almost as big as Sojourner itself.

Mars robot, Sojourner, JPL, USA

Small Rover with Folding Wheels

Sojourner, however, was not the only project developed within the scope of the JPL's Rover and Telerobotics Technology program. Led by Dr. Paul Schenker, the research team was responsible for the Lightweight and Survivable Rover (LSR) program in 1998 and currently works on a whole range of mobile robots intended to explore other planets.

The rover, LSR-1, is a research prototype weighing just 7 kilograms. Its wheels have a diameter of 20 centimeters and are foldable. The vehicle is 97 centimeters long, 70 centimeters wide, and has a ground clearance of 29 centimeters. In comparison, Sojourner weighed over 11 kilograms, was 63 centimeters long and just 45 centimeters wide. It also had a ground clearance 7 centimeters smaller. LSR-1 can overcome obstacles up to 40 centimeters high, can stay operational for several months despite extreme fluctuations in temperature, and can travel several kilometers in the process.

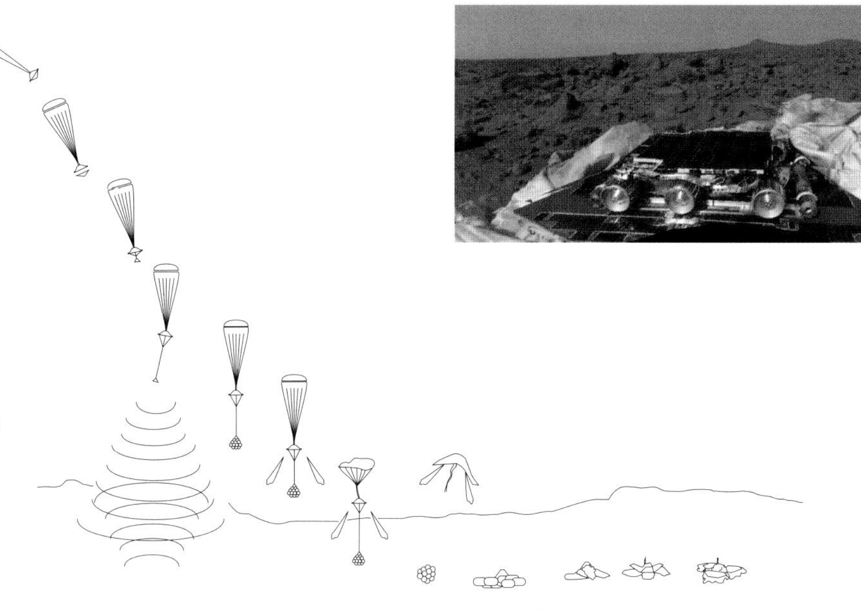

Geology on Mars

However, the NASA's Mars missions are not just about small autonomous vehicles. In the January of 1999, MVACS (Mars Volatiles and Climate Surveyor) began a mission to explore geological and climatic conditions on Mars. NASA is searching for water and carbon dioxide on the red planet. It is also planned that the ground temperature and thermal conductivity of the red soil will be measured.

In order to do this, the scientists integrated a robotic arm, Mars Arm II, designed by Dr. Hari Das, into the Mars landing craft. This arm possesses four degrees of freedom and is affixed to the upper surface of the "Lander". Attached to the last segment is a sort of heated digger scoop with a tube and integrated thermoelement affixed to the back of it. Using the robotic arm, the tube can be inserted into the ground and the temperature measured several centimeters below the surface. If the tube is heated up and the time required for the tube to cool down is again measured, other geological properties of the soil can be concluded.

The digger scoop has the function of taking soil samples and placing them in a measuring chamber on the Lander's upper deck. By heating up the scoop, the process of filling the chamber should be facilitated, as particles which have stuck or frozen onto the scoop are unable to detach on their own. Among other things, CO_2 and water content will be determined in the measuring chamber. The landing craft itself possesses sensor equipment for measuring atmospheric parameters such as wind, temperature and pressure.

Planet rover LSR-1 together with Sojourner, JPL, USA

Planet rover LSR-1 with the robotic arm Micro Arm 1, JPL, USA

If a mission sends another rover to Mars or to the Moon that must take samples far away from the landing craft, the JPL engineers are now prepared. The LSR-1 is equipped to carry the small robotic arm called Micro Arm I, which is capable of performing tasks similar to those of Mars Arm II on open terrain.

Multisensory Space Robotics from Germany

The German Space Center (Deutsche Forschungsanstalt für Luft- und Raumfahrt DLR), in Oberpfaffenhofen, Germany, pursues a holistic method in developing robots. When designing this new generation of robots, it integrates lightweight construction, multisensory intelligence, localized autonomy, remote-controllability and self-teaching abilities from the start on a broad basis.

During the mission Spacelab-D2, a milestone in the history of space travel was reached in April 1993. For the first time, Rotex, a small robot equipped with localized, "multisensory" intelligence, carried out tasks flexibly in a wide range of operating modes on board a spaceship. It was pre-programmed (and re-programmed from the ground during the mission) by astronauts using the DLR control ball, and remote-controlled using a stereo TV monitor, and also teleprogrammed and remote-controlled from the ground.

In these modes of operation, the robot had to disconnect and reconnect plugs with a bayonet-like fitting, assemble and dismantle mechanical structures, and catch a free-flying object.

- **Rotex carrying out internal servicing on the Spacelab-D2 mission, DLR, Germany**
- **Internal space robot, DASA and DLR, Germany**

Graphic Simulation Thinks Ahead

Based on onboard sensor feedback, the localized autonomy is responsible for the "shared autonomy". In this way, in a normal situation, the robot tries independently to improve the rough incoming commands from human operators or from its path planner by using its sensors. The only exception (and particularly spectacular because of the difficulty involved) was catching a free-flying object fully autonomously, controlled solely by the image-processing computer on the ground, and with a six-second signal transit time.

To compensate for the signal delay time, the robot's behavior had to be calculated in advance under the influence of sensory perception and localized signal feedback as well. A predictive, three-dimensional graphic simulation was used for this, which compensated for the delay time. In this way, the operator together with the computational intelligence on the ground enabled not only the sensor-assisted teleprogramming, but the sensor-assisted online remote control, as well.

The knowledge acquired from Rotex now forms the basis for future space travel robots. The methods implemented here can be used directly for internal and external servicing.

An Assistant for Astronauts

Rotex and the other developments that have been carried out to date have enabled later experiments in space labs to build upon the knowledge gained from them. It could even be that the next big step will be an operational system for assisting astronauts or, to some extent, even replacing them. It should either be mobile on three linear tracks or take the form of a climbing system.

If appropriately consistent progress is made in the research and development work, this type of system could be largely autonomous and remote-controlled. The implementation of robots on externally mounted experiment panels is equally well-prepared for as the internal conditions for experiments.

signal transit times

Electromagnetic waves travel at the speed of light ($3 \cdot 10^8$ m/s). However, the signal's transmission takes several seconds, due not only to the distance from the Earth to the orbiting satellite and back again but also to the number of relay stations involved in the signal's pathway. In the SkyLab mission, for example, 20 relay stations were needed between Oberpfaffenhofen, Germany, SkyLab, and back to transmit the signal. The computers belonging to these relay stations had to check the data transmission records, delaying the total transmission time – in this case, approximately six seconds.

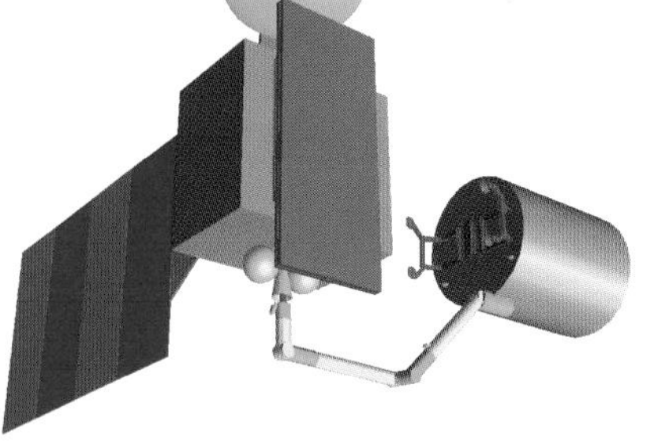

ESS docking at the apogee motor of a defective TV satellite, DASA/DLR, Germany

ERA robotic arm with ERA extension (yellow), DASA, Germany

Reviving the TV satellite?

The ESS (Experimental Servicing Satellite) project, initiated by DaimlerChrysler Aerospace (DASA), is aimed at the preliminary development of free-flying telerobots. They fly to space systems to inspect them and carry out maintenance.

The TV satellite, TV-Sat 1, which was defective due to the fact that one of the solar panels had not opened after its positioning in orbit, was selected as a demonstration object for one of the first experimental systems. Even today this type of geostationary news satellite has still not been designed for maintenance to be carried out by robot; but in this instance it was possible to use the control jet of its apogee motor as a natural docking port.

apogee motor:
An apogee motor is a rocket engine with which every geostationary satellite must be equipped. The carrier rocket only brings the satellite into an elliptical orbit around the Earth. The distance of this ellipse at its point furthest from the Earth (the apogee) is 36,000 kilometers. The apogee motor accelerates the satellite away from this point in such a way that it can reach the circular geostationary orbit.

In the ESS project, a new type of capture tool was developed which enabled the robot's end-effector to be completely inserted into the apogee jet. To do this, it possessed a power sensor, two arrays placed longitudinally on the tool each with three laser distance-measuring devices arranged in a star-shape, a TV camera, and a specially adapted splaying mechanism.

On reaching its objective, the capture tool drew in the satellite in need of repair and clasped it with a simple gripping mechanism. When the arm was free again, by way of an interchangeable adapter, it picked up a pair of electromechanical clippers, which had also been newly developed during the project, and cut the clamps blocking the TV satellite's solar panel.

ESS capture tool – docking at the apogee motor, DASA/DLR, Germany

Mechanic in Space

Commissioned by the European Space Agency ESA, the European robotic arm ERA is currently being built, integrated, and tested by, among others, the companies Fokker, DASA-RI, and TS. ERA is intended to be used for assembly and maintenance work on the Russian segment of the International Space Station (ISS) and should be ready for use in space by the year 2000 or 2001.

ERA is a robotic arm approximately 11 meters in length possessing seven degrees of freedom which have been constructed symmetrically. It can move around the space station independently and can be affixed to defined docking adapters. For future developments (ERA extension), it is intended that the arm will be equipped with additional smaller robotic arms (approximately 1.5 meters long) to enable it to carry out more complex repairs on the space station.

The Astronaut's Third Hand

To assist astronauts inside the space station, stationary or mobile robots will be introduced for operating the individual scientific experiments. Here, the ability of the robot's kinematics to adapt to suit various task requirements is particularly important. Also, the robotic arm should weigh as little as possible but be able to carry heavy weights.

Accordingly, developments are being carried out by DaimlerChrysler Aerospace and DLR, Germany, at the moment. One example is a mobile robot possessing three arms approximately 100 centimeters in length, which it uses to move along the instrument lockers and to operate the individual experiments.

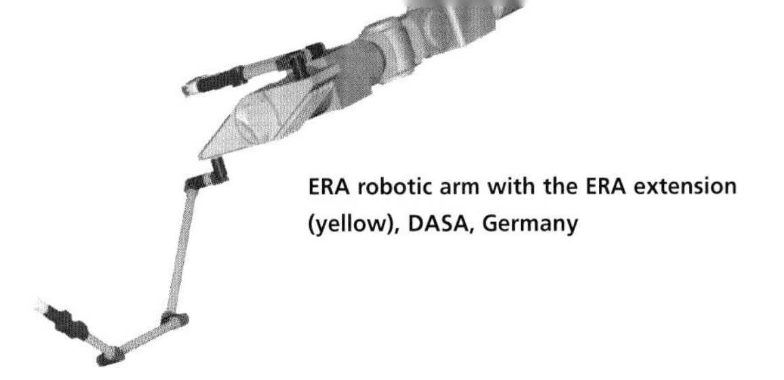

ERA robotic arm with the ERA extension (yellow), DASA, Germany

Flying Helper

Inspector is a free-flying service system made by DaimlerChrysler Aerospace. It is intended to be implemented in the planned International Space Station (ISS) to assist astronauts when they go for walks in space, to monitor places that are difficult to reach, and to perform repair and maintenance tasks.

In the preliminary stage (X-Mir Inspector), the service system has been equipped with a new type of video system for monitoring and inspection work, and has already been tested on the Russian space station, Mir. It is planned that later on in the development, the service system will be equipped with additional manipulator arms, allowing it to dock at the space station and carry out maintenance and repair tasks.

Within the scope of the International Space Station, it is planned that panels will be installed on the latticework structure outside the pressurized module in order to set up apparatuses for carrying out scientific experiments. The operation and care of these experiments will be performed by a TEF (Technology Exposure Facility). TEF is a small, externally stationed experiment robot with arms between 1.5 and 2 meters in length that possess seven degrees of freedom and which are equipped with the appropriate external sensors, such as force momentum sensors, CCD cameras, and laser diodes.

Spin-off for Civilian Use

Cost and investment returns are not the main issue in the prestigious world of space travel. However, this field is of particular interest in the development of robotics technology because space robotic developments will soon enrich earthly service robot developments, where practical use comes before fascination.

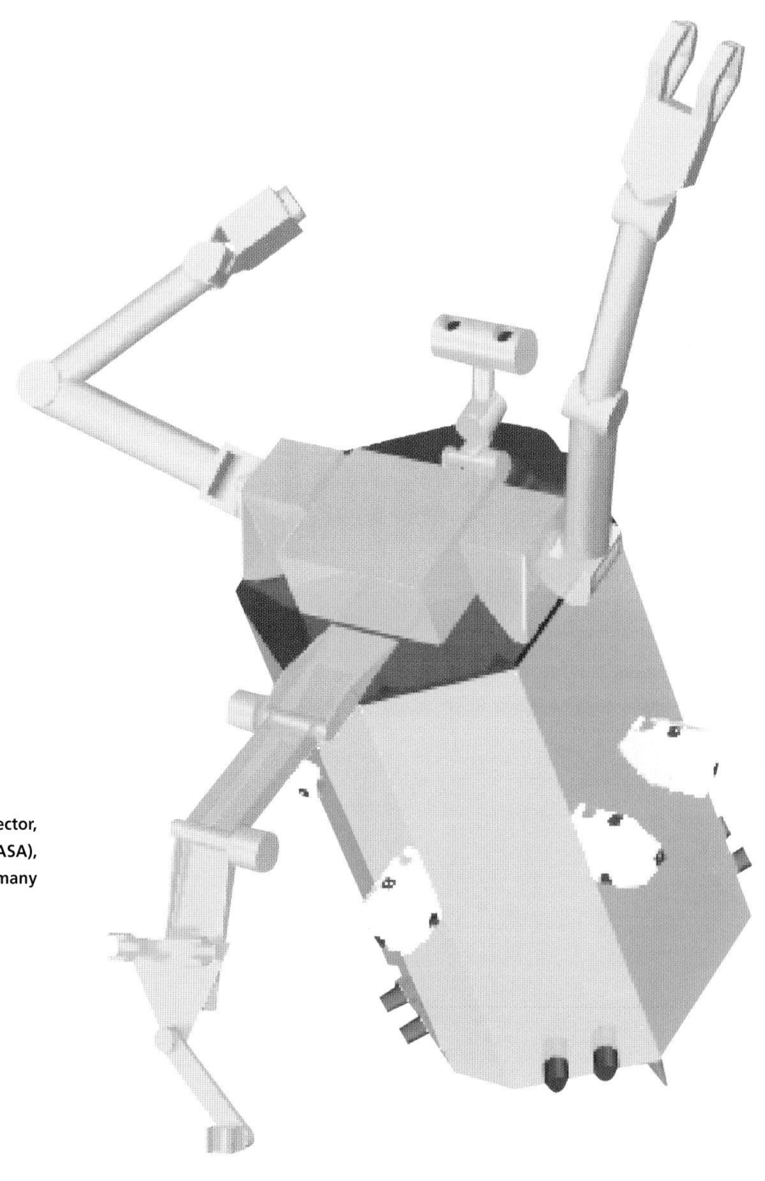

Free-flying service system, Inspector, DaimlerChrysler Aerospace (DASA), Germany

Future

Preliminary sketches

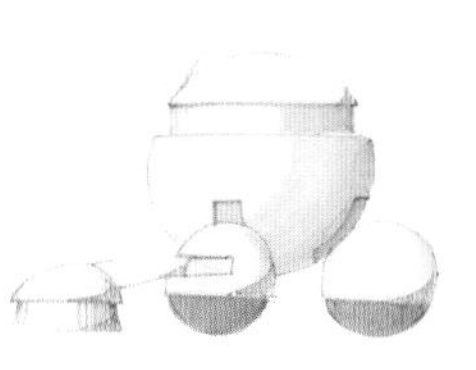

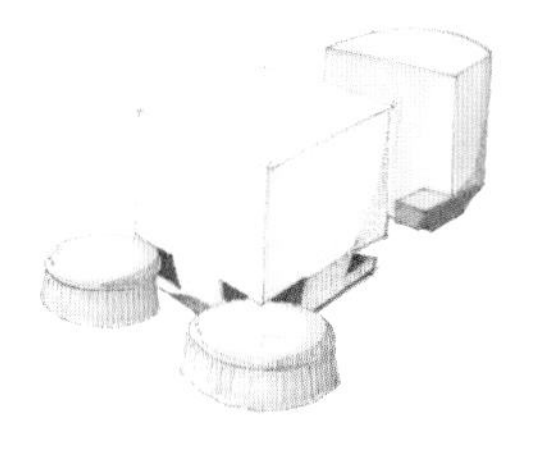

Service robot for cleaning streets and squares, Lothar Kotulla and Andreas Kull, HfG Schwäbisch Gmünd, Germany

Service Robots in Everyday Life

What is in store for service robots in the future? What applications will we see? The creativity and imagination of engineers the world over are called upon to answer these questions.

In this closing chapter, we illustrate the infinite number of new fields of implementation by describing three scenarios involving service robots that could become part of our daily life within the next three, five, or ten years. The visions portrayed here have come from engineers confronted every day with robotic systems, as well as product designers who have concerned themselves with robots ever since the first science fiction films were made.

Robot in Town

In 1997, two students, Andreas Kull and Lothar Kotulla, doing their theses at the Hochschule für Gestaltung FH, Schwäbisch Gmünd, Germany, designed an unmanned cleaning machine for use in cleaning public streets and squares. The technical concept was then developed in cooperation with Fraunhofer IPA.

The autonomous working robot acts as a mechanical assistant for cleaning personnel. The human workers are no longer bound to the machine; they can work freely and flexibly with the robot. Man and machine make for an efficient cleaning team. The cleaning personnel do, however, have the option of manual control. The concept is maintenance-friendly and planned on a modular construction with detachable drive and function modules, such as with lawn mowers and road-paving machines.

The robot's motion paths are programmed in teach-in mode. A digital town map provides orientation and the robot can find its way around on it by using a satellite navigation system.

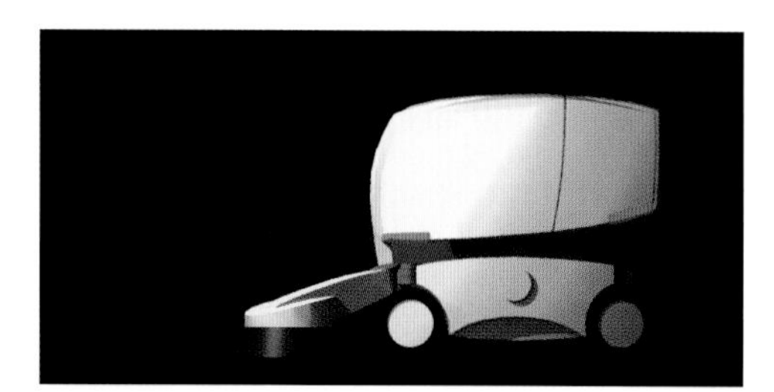

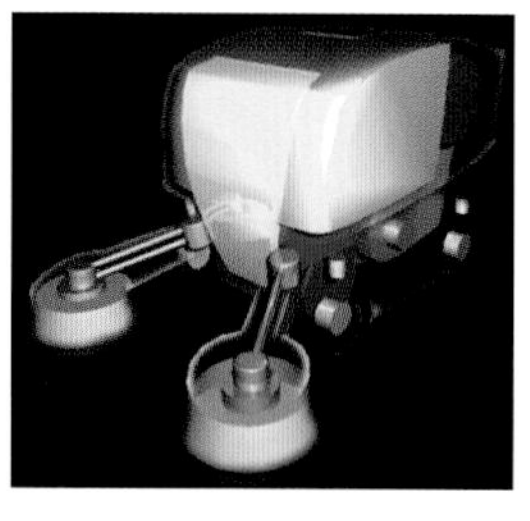

Representation of the service robot's technical components

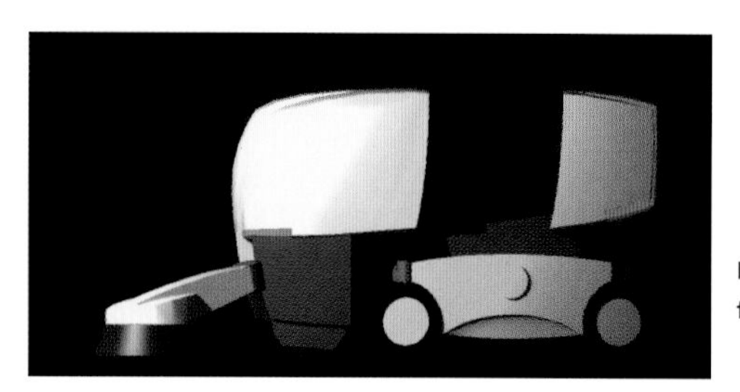

Detaching the function module from the motion platform

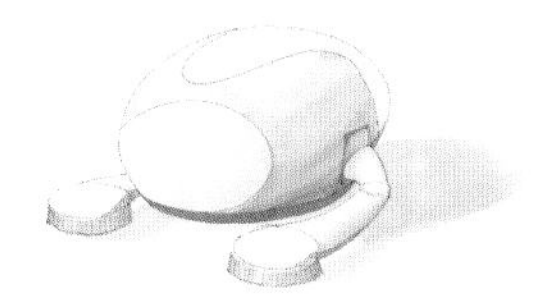

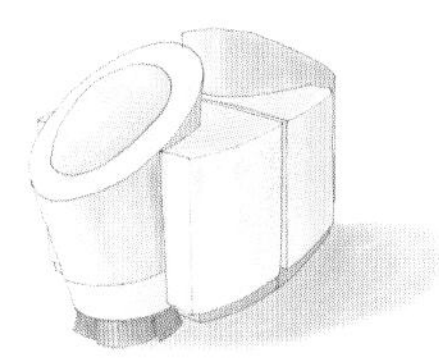

Radar sensors identify obstacles, and acoustic and visual signals indicate the robot's proper reaction, such as stopping or swerving.

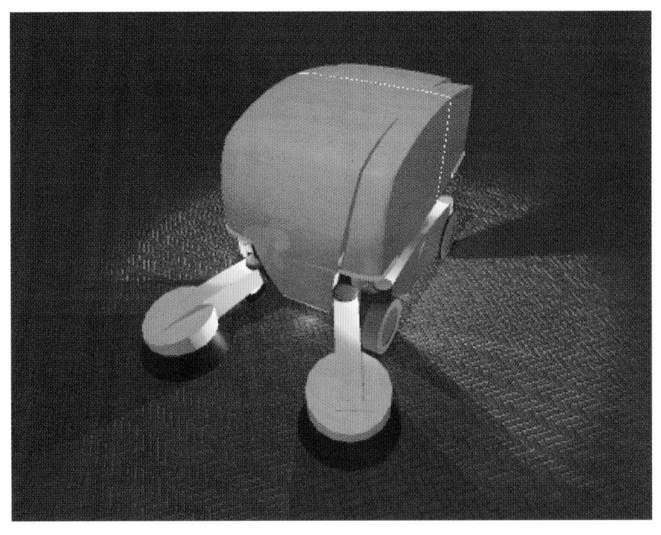

Using light to demarcate the working area

A Robot with Distinctive Looks

To make people aware of the robot's presence, it has been equipped with a strip of flashing lights. Additional lights create a visual barrier on the ground to demarcate the working area. The type of lighting should function as a characterizing feature of the robot to differentiate it from other machines.

Large, domed surfaces create an attractive appearance. Technical, angular shapes serve to maintain a machine's character; the robot must always remain recognizable as a working machine. A friendly and dependable appearance is particularly important for a robot as it should not be seen by passersby as a threat, but instead as an accepted part of future street life.

If such a robot were always working downtown, it would carry prestige for clean and progressive towns, larger trading areas, and industrial enterprises.

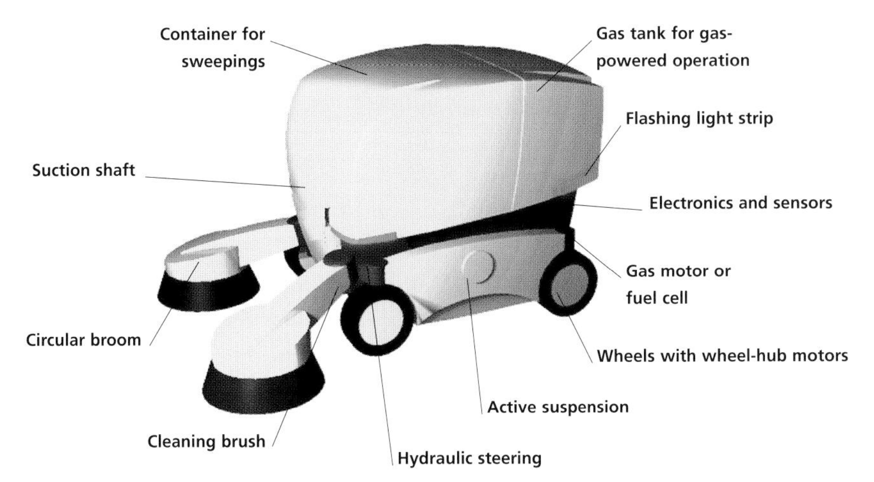

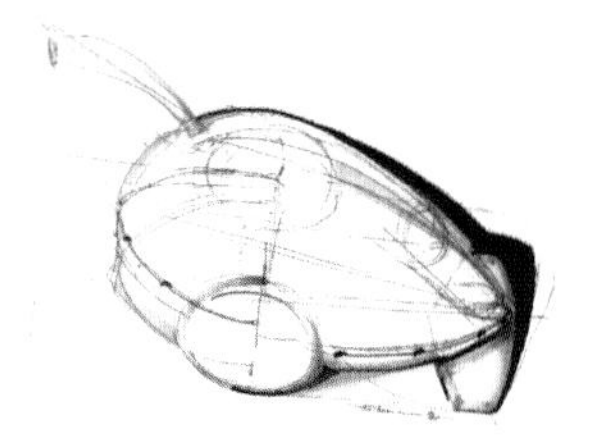

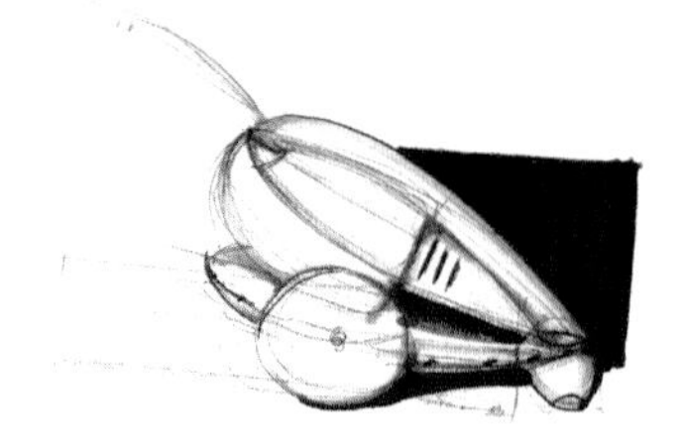

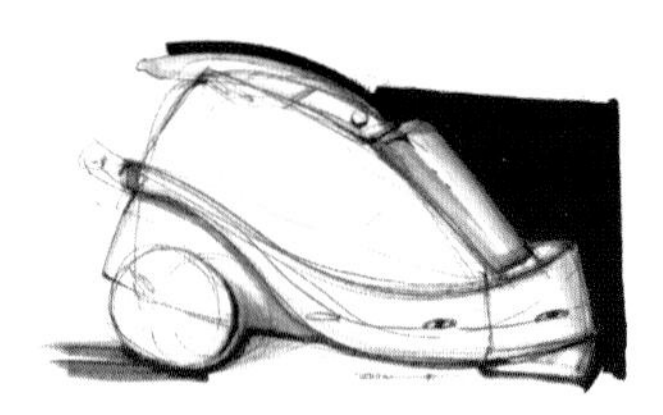

Design thesis on an autonomous vacuum cleaner, Jochen Bittermann, HfG Offenbach, Germany

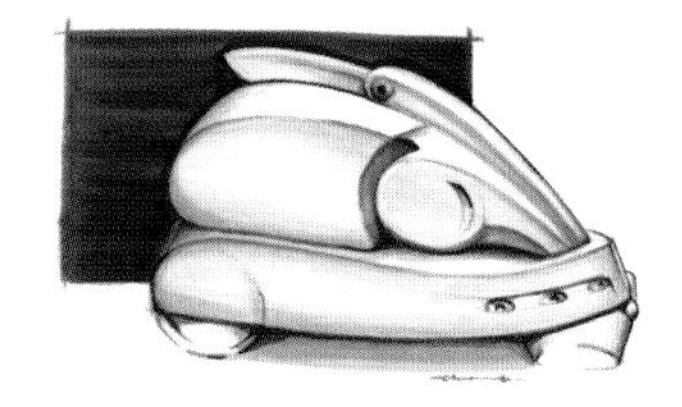

Robots that Tidy Up

In the home of the future, an autonomously functioning vacuum cleaner could relieve housekeepers of the monotonous job of floor cleaning. Its independent behavior and spatial orientation would give the observer the impression that he is housing a being in his home with its own life. Although the vacuum cleaning robot is only equipped with simple sensor units, it has some of the charm of a household pet. The vacuum cleaning "creature" is intended to represent familiarity, efficiency, and reliability. Only then would it be accepted in private households.

The work done by Jochen Bittermann at the Hochschule für Gestaltung in Offenbach, Germany, is principally concerned with the design of the service robot Skarabäus and its integration into a dynamic home environment. Skarabäus vacuums the room constantly and independently but can be corrected and controlled by its owner at any time. If a glass of red wine has been left standing on the carpet, the robot maneuvers gently around it.

Using a 180° ultrasound sensor unit and tactile sensors fixed onto a fender strip, Skarabäus is able to orient itself in space horizontally and, to some extent, vertically as well; by using its spatial knowledge, it adapts itself to its environment.

Manual Control on Stairs

Obstacles such as steps and stairs can not yet be overcome by a cost-effective cleaning robot. A compromise here is manual intervention: the integrated hand vacuum cleaner is detached from the vehicle platform and steps, window sills, and recently contaminated surfaces are vacuumed by hand.

All the cleaning functions are hidden in the hand vacuum cleaner. The vehicle's mobile platform orients and navigates. If the hand vacuum cleaner is reinstalled onto the vehicle platform, the components become a closed unit again. The vehicle platform now functions as a recharging station for the hand vacuum cleaner. Now the robot has become autonomously active again and carries on vacuuming in the room.

If the power of the vehicle platform's integrated batteries is low, Skarabäus aims for the docking station, which is not only its recharging station but also its "home". The station, which is half covered by a roof, is somewhat reminiscent of a doghouse. It has an integrated touchscreen display for entering cleaning times and displaying operating instructions, such as "change dust bag".

Multitalented Helper in the Supermarket

In Germany, gas station shops and modern train stations with shopping centers serve customers by allowing them to shop while on the go. In 1995, 13% of the turnover in the food retail trade took place in gas station shops, and 40% of gas station customers only went there to shop!

How can conventional supermarkets fight against this substantial loss of turnover taking into account existing German shop-trading laws that prohibit a "24 hours 7 days a week" service? A study carried out by Fraunhofer IPA showed that the night could be put to good use in supermarkets using service robots, whilst at the same time offering unlimited shopping possibilities.

A service robot with a visual recognition system and a bar code scanner can fill up a shelf in a shop with articles from stock. Its grippers must be able to hold any package gently but securely. To cope with the wide range of shapes and surface structures of supermarket articles, the gripper possesses various systems for force-controlled handling: even a carton of eggs is safe in its hands.

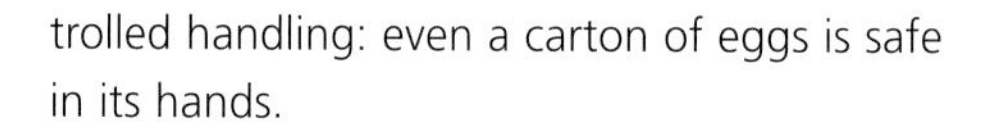

The mobile, autonomous, freely moving vehicle requires navigational abilities and avoidance strategies in order to find its way around the supermarket aisles and to react appropriately when confronted by obstacles. It needs to be able to communicate so that it can coordinate its tasks with a control point.

When updating inventories, the service robot is capable of checking stocks for inventory and expiration dates, recognizing damage and misplacement by customers, and fetching replacement items from the warehouse. With the appropriate functional units, cleaning tasks, which are performed more easily in an empty shop, could also be carried out.

Individual Customer Service Around the Clock

Combined with a touchscreen order terminal, credit card scanner, and a lock for the wares at the front of the shop, articles could also be sold automatically during the night. With the same functionality, remote ordering could also be carried out on the Internet and these items could then be sent out by workers the following morning. The central computer would take care of the chronological coordination of all the tasks according to priority.

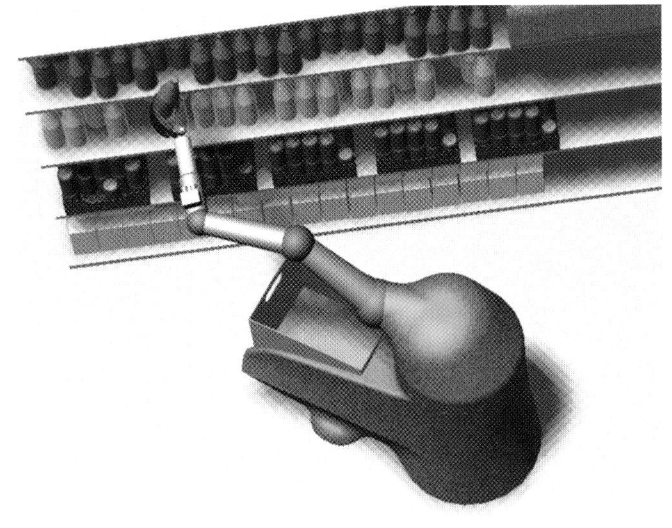

Supermarket robot, Fraunhofer IPA, Germany

For this scenario to be feasible, the service robot would need to be able to be integrated into an existing infrastructure that is operated manually rather than automatically. Of course, the robot could be used for many other applications, in addition to supermarkets.

Any doubts concerning the safety aspect of a fully unsupervised system could be overcome using a central remote monitoring system. With the appropriate equipment, such as a camera and tear gas, the service robot could be activated to ward off intruders. Its versatility makes such a multitalented robot a profitable investment.

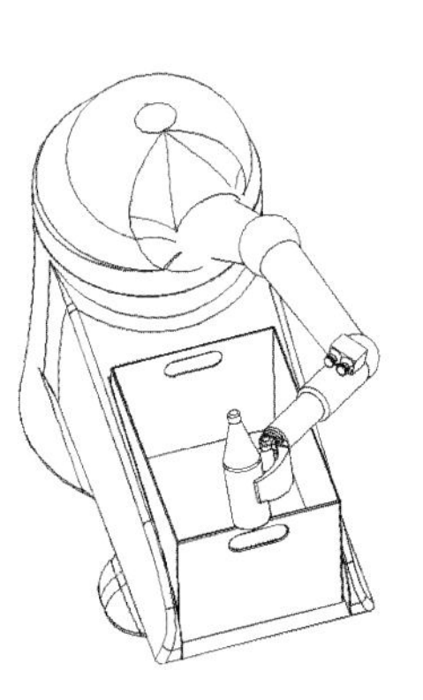

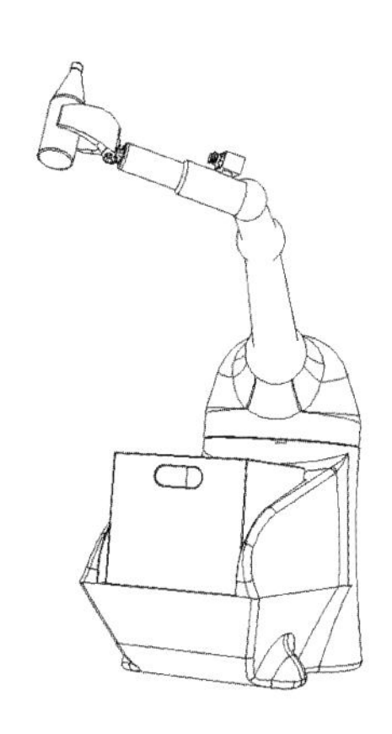

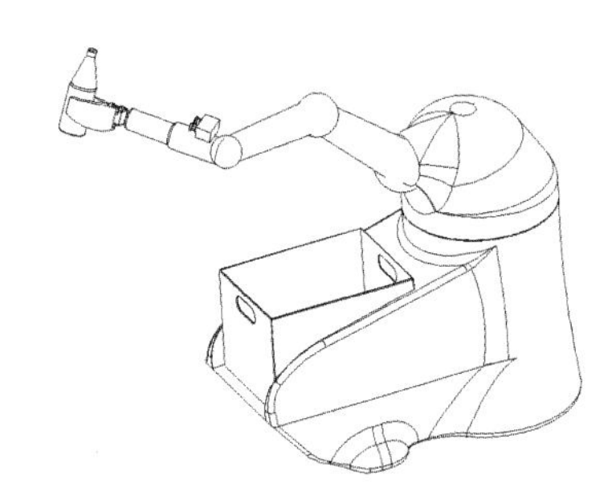

Due to the breakneck speed of developments in the field of service robots, this book will inevitably lose its up-to-dateness.

Instead of a static list of references, we therefore refer you to a WWW page that will actively keep pace with the latest service robot developments.

enter

http://www.ipa.fhg.de/srdatabase

CLOSE